MECHANICS OF HEARING

MECHANICS OF HEARING

PROCEEDINGS of the IUTAM/ICA SYMPOSIUM

held at

Delft University of Technology
The Netherlands
13-15 July 1983

Edited by E. de Boer and M.A. Viergever

1983 MARTINUS NIJHOFF PUBLISHERS
a member of the KLUWER ACADEMIC PUBLISHERS GROUP
THE HAGUE / BOSTON / LANCASTER

and

DELFT UNIVERSITY PRESS
DELFT

Distributors

for the United States and Canada:
Kluwer Boston, Inc.
190 Old Derby Street
Hingham, MA 02043
USA

for all other countries:
Kluwer Academic Publishers Group
Distribution Centre
P.O. Box 322
3300 AH Dordrecht
The Netherlands

Cover

Front cover shows a painting, entitled 'Auricles', 1982,
by Inez Merhottein, Voorburg, The Netherlands.

Book information

Joint edition published by:
Martinus Nijhoff Publishers, The Hague,
The Netherlands
and
Delft University Press, Delft, The Netherlands

ISBN-13: 978-94-009-6913-1   e-ISBN-13: 978-94-009-6911-7
DOI: 10.1007/978-94-009-6911-7

PROSPERO
– Dost thou hear?
MIRANDA
– Your tale, Sir, would
  cure deafness.

Shakespeare,
The Tempest, Act 1, Sc. 2.

CONTENTS

Welcome to the readers

This book gives an account of present-day attempts at solving the problems
posed by the truly amazing capabilities or our hearing organs. The emphasis
is on those aspects of the external ear, the middle ear and the cochlea which,
to the best of our present knowledge, can be treated by a mechanistic analysis. The book represents the proceedings of a Symposium on Mechanics of Hearing, held at Delft (the Netherlands) in July 1983. The symposium was jointly
sponsored by the International Union of Theoretical and Applied Mechanics
(IUTAM) and the International Commission on Acoustics (ICA) and it functioned
as a special symposium associated with the 11th International Congress on
Acoustics in Paris.

A scientific committee was appointed (see list below) under the chairmanship
of the undersigned. The committee selected a number of possible contributors,
and requested suggestions for additional contributors. In this way the core
of the symposium programme was constructed. Each author had to produce a
camera-ready manuscript which means that the authors are fully responsible
for their texts. In a few instances the Bureau of the Symposium provided
help to the authors to ensure that all manuscripts were typed according to
the same rules. The book was made available at the time of the Symposium
thanks to the diligence of Delft University Press.

The following gives a guide as to the contents of the book. The first topic
is called: 'External ear and middle ear'. A review paper by Shaw and Stinson
analyzes the many physical properties that have been demonstrated in these
organs. Sound is affected by structures of complex geometry. A modern way of
attacking the problem of complexity is demonstrated by Funnell in his contribution on vibrations of the drum membrane. Sound does not only go from 'air'
to 'ear' but - in view of cochlear emissions and other active, nonlinear phenomena - also in the opposite direction. See the paper by Matthews on the
transmission of sound generated in the inner ear back to the middle ear and
to the external ear.

The second main section of the book concerns 'Cochlear fluid mechanics', this subject more or less represents the 'classical' approach in cochlear mechanics. The section brings together a number of papers, mostly of a fundamental nature, treating the problem as to how cochlear fluids interact with cochlear membranes. One of the most versatile solution methods, the LG (Liouville-Green) or WKB (Wentzel, Kramers and Brillouin) method, is applied by Steele and Zais to cochlear structures of fairly complex geometry. Other applications of that method in two- and three-channel cochlear models, are presented by Babič and Novoselova. A different approach, more easily recognized as an asymptotic method, is illustrated by Holmes and Cole. To what extent experimental data can be explained in terms of 'classical' models is demonstrated by Viergever and Diependaal. Very fundamental properties of models incorporating 'short waves' and the ways these properties are interconnected form the topic of the paper by Lighthill.

With the advent of 'Cochlear emissions' - the name of the next section - a new era seemed to start in the field of hearing theory. New experimental findings and a novel interpretation are presented in Kemp's contribution to this book. It is difficult to grasp all data and to construct a comprehensive model to explain them all, hence the studies of simplified models. See, for instance, the paper by Sutton and Wilson. More experimental data are presented by Rutten and Buisman. These authors also relate the emission phenomena to subjects of study in completely different fields of research: phase transitions of oscillators that operate near their critical points. One bridge too far? Certainly not! Finally, Wit and Ritsma consider spontaneous emissions in frequency and time. They also try to determine the minimal stimulus level that gives rise to an evoked emission - with a surprising result.

The fourth main topic, obviously related to the previous one, is 'Active Systems'. Several authors in this field claim that the classical, passive cochlea model is not capable of explaining the essential elements of the most recent findings regarding vibrations in the cochlea. When a model is assumed to be active, i.e., to have the property that cochlear structures can actively generate acoustic energy, it is feasible to obtain a well-fitting response. This is demonstrated by the model responses obtained by Neely. Mountain, Hubbard and McMullen describe more general aspects of an active model and the way computed responses relate to experimental evidence, whereas Koshigoe and Tubis concentrate on feedback properties of an active model.

In both papers nonlinearity appears as an essential feature of the model. Problems associated with reflections of waves generated inside the cochlea by active behaviour are analyzed by de Boer. Van Netten and Duifhuis give an account of their first attempt at an analytical approach: the dynamics of the organ of Corti is described by the Van der Pol equation. Diependaal and Viergever studied numerical techniques for solving the problem of an active structure. They find one of the most advanced methods to fall definitely short of the goal in active models and they elucidate the reason for this property.

The fifth section of the book, entitled 'Nonlinear micromechanics', tries to delve somewhat deeper into the problem of how the specific properties of the organ of Corti are brought about. Voldrich presents an account of the most recent anatomical findings. Jau and Geisler consider nonlinear effects as dependent upon a weighted average of basilar membrane displacement over a certain length. Khanna and Leonard enumerate the arguments why they think tuning properties of the cilia of cochlear hair cells are crucial, a feature that has been keeping theorists busy for a long time.

A number of 'Special topics' remain, difficult to be brought under one heading. Miller's experimental results on the static compliance of the basilar membrane contribute to present-day discussion on this topic. Allen shows how the dynamics of neural excitation in hair cells can be taken into account. Questions of sound conduction in water birds, the role of the cochlear aqueduct and the significance of a flexible spiral lamina are considered by Kohlloeffel. Bialek, finally, approaches the problem of cochlear action from quite a different angle. He calculates the noise level resulting from Brownian motion and finds this to be at least 20 - 30 dB above our hearing threshold. According to these results there should exist a filtering process subsequent to cochlear mechanics.

Many thanks are due to the sponsoring agencies: IUTAM and ICA. The cooperation with the Department of Mathematics and Informatics, Delft University of Technology, served well to make the meeting a success. The undersigned gratefully acknowledges the work done by the scientific committee (see list below). The many contributions from Dr. M.A. Viergever who assumed the laborious task of Secretary of the Symposium, deserve to be mentioned specifically.

The same applies to Mrs. M. den Boef who undertook the greater part of the local organization and administration work.

Let us all hope that the present book will be a milestone along the road of modelling of the auditory system.

E. de Boer
chairman

The scientific Committee:

Sir James Lighthill, London, UK
C.R. Steele, Stanford, CA, USA
E.A.G. Shaw, Ottawa, Canada
E. de Boer, Amsterdam, Neth. (chairman)
M.A. Viergever, Delft, Neth. (secretary)

Section I

External ear and middle ear

THE HUMAN EXTERNAL AND MIDDLE EAR: MODELS AND CONCEPTS

E.A.G. Shaw, M.R. Stinson

National Research Council
Ottawa, Canada

ABSTRACT

The performance of the external ear, when viewed as a diffuse-field receiver, is given in a simple expression containing two acoustic impedances. In this sense, the external ear has a high frequency performance quite close to the theoretical limit. Viewed as a directional antenna the external ear is an acoustical wave processor of considerable complexity. Approximately eight normal modes spread over nearly three octaves are required to account for its distinctive characteristics. At the highest frequencies, additional wave factors come into play near the eardrum. Network concepts are well suited to the mechanics of the middle ear but require considerable development to allow for the complex motion of the eardrum which is the dominant factor at high frequencies. Considerable progress has been made with a two-piston model which gives reasonable eardrum impedance and middle ear transmission curves. This model shows that, at high frequencies, it is the internal resistance of the eardrum that absorbs most of the incident sound energy and controls middle ear transmission. A more sophisticated treatment of eardrum motion may soon be within reach.

1.   RECEPTION IN A DIFFUSE SOUND FIELD

The primary function of the external ear, the collection of acoustical energy, can be quantified in a precise manner by performing a mental experiment in which the ear is first a receiver and then a transmitter. Hence, by invoking the acoustical reciprocity theorem, it can be shown (Shaw 1979) that the power P_d absorbed at the eardrum, when the ear is immersed in a diffuse sound field of mean square pressure p_f^2, is determined in essence by two impedances as follows:

$$P_d = (\lambda^2/4\pi)\left[4\eta R_a R_d/\ |Z_a+Z_d|^2\right] p_f^2/\rho c \qquad (1)$$

In this expression, Z_a is the acoustic impedance seen by the eardrum looking outward through the external ear, Z_d is the impedance presented to the external ear by the middle ear system, R_a and R_d are the resistive parts of these impedances, and λ is the wavelength of sound. When Z_a is the conjugate of Z_d and when the radiation efficiency η is 100% (no sound absorption between the eardrum and the diffuse field), the power received at the eardrum has its greatest possible value

$$P = (\lambda^2/4\pi)(p_f^2/\rho c). \qquad (2)$$

This is the total power flowing through a transparent sphere of cross-sectional area $\lambda^2/4\pi$ (radius $\lambda/2\pi$) when immersed in the same diffuse sound

4

field. Similarly, comparing equations (1) and (2), it follows that the
"absorption cross section" of the non-ideal ear is

$$A = (\lambda^2/4\pi)\left[4\eta R_a R_d / \; Z_a + Z_d \; {}^2\right] \tag{3}$$

The upper solid curve in Fig. 1 shows the calculated absorption cross section
for a physical model of the external ear whose acoustical characteristics are
closely matched to those of the median human ear (Shaw 1982). The values of
Z_a for this model were obtained from impedance tube measurements and the
values of Z_d came from a middle ear network (see Shaw 1982). The validity of
the method was confirmed by measurements of diffuse field response (Shaw
1979). As can be seen, the human external ear is a poor sound collector at
low frequencies but approaches the theoretical limit of performance at its
principal resonance frequency ($\sim$ 2.7 kHz) and at higher frequencies. For
comparison, the broken line in Fig. 1 shows the calculated absorption cross
section when the external ear is eliminated and the eardrum is placed on the
surface of a sphere representing the head.

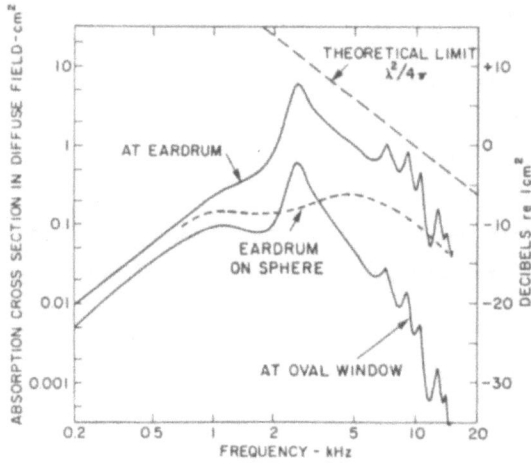

*Fig. 1. Calculated absorption
cross section for external ear
model combined with middle ear
network. Difference between
solid curves indicates fraction
of absorbed energy reaching
cochlea. Broken line shows
absorption cross section with
external ear eliminated.
(Adapted from Shaw 1982.)*

2.   DIRECTIONALITY AND NORMAL MODES

Measurements indicate that the diffuse field response of the human ear is
relatively insensitive to variations in geometry (Shaw 1980). The directionality of the ear, which is highly significant in sound localization, is
however closely linked with its geometry as shown in Fig. 2. These families
of response curves were obtained with a special source designed to produce
clean progressive waves at grazing incidence. The measurements were made at

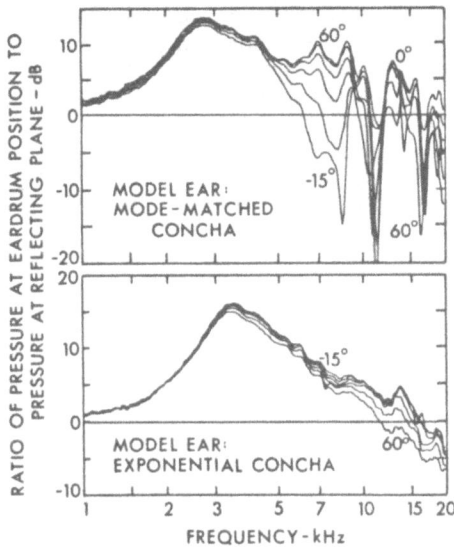

Fig. 2. Frequency response at six angles of incidence. Upper panel: Model ear with concha acoustically matched to median human ear. Lower panel: Model ear with exponential concha.

six source positions simulating median plane excitation at elevations betwen -15° and +60°. It is evident that the response of the model ear with "human" acoustical characteristics (upper panel) is highly directional at frequencies greater than 5 kHz whereas the response of the ear with the exponential concha is substantially independent of source position at all frequencies.

The key to the directionality of the human ear is found in the normal modes of the concha (e.g. Shaw 1982). The pressure distributions symbolized in Fig. 3 are based on measurements on ten human ears with the ear canal closed. They are presented against the geometrical representation of the concha used in the model ear referred to in Fig. 2. For each measurement, the source position was carefully selected to excite the chosen mode while minimizing the excitation of adjacent modes. The resonance frequencies shown are the mean values for the ten subjects and the arrows indicate the directions of maximum excitation. These frequencies and directions are almost perfectly matched in the model ear. Notice that the second and third modes have

Fig. 3. Relative phases of sound pressure in different parts of concha based on data from ten human subjects. Sketch at left outlines cavity system in human concha.

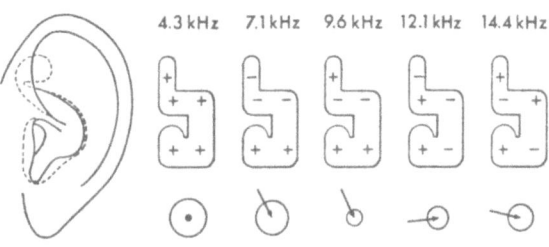

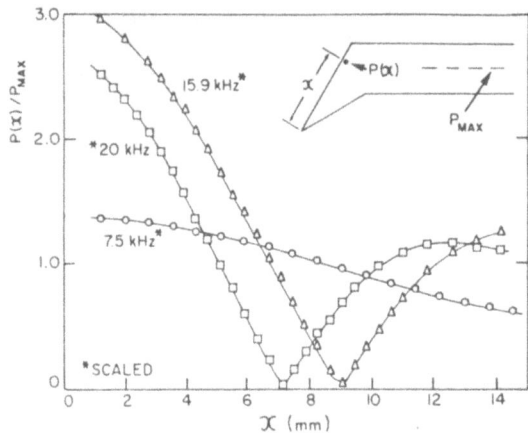

Fig. 4. Measured pressure distributions in physical model representing human ear canal with bend and tapered end. P(x): Pressure magnitude along x-axis. P_{max}: Magnitude of pressure maxima in uniform cylinder.

pressure distributions that are primarily vertical while the fourth and fifth modes are predominantly horizontal. As indicated in Figs. 1 and 2, the doublets become triplets in the complete ear (canal open).

3. EAR CANAL GEOMETRY

It has been customary to treat the human ear canal as a uniform cylindrical tube terminated by a plane eardrum set perpendicular to the canal axis. In reality, the eardrum is inclined to the canal axis forming a wedge-shaped volume at the end of the canal which tends to be bent away from the main body of the canal. The cavity depicted in Fig. 4 is one of an experimental series representing such characteristics (Stinson and Shaw 1982). As can be seen, the measured pressure distributions are significantly different from those associated with the simple canal. In particular, at the highest frequencies, the pressure maxima in the tapered portion of the cavity are much greater than those in the uniform cylindrical region. Furthermore, the first zone of the wave pattern is considerably extended which places the first minimum at a distance of approximately 0.4λ from the end rather than 0.25λ. The benefit conferred by this horn-like behaviour is offset by the presence of a pressure node across the inclined eardrum surface.

4. MIDDLE EAR MECHANICS

For more than two decades lumped-element modelling has provided a valuable framework for middle ear mechanics. In particular, Zwislocki's well known acoustical network (Zwislocki 1962) gives a good account of normal and pathological human middle ears at frequencies up to 1 or 2 kHz, as judged by the quality of fit between the calculated and measured values of input

*Fig. 5. Compound-eardrum
concept (a) S_o: rigidly
attached to malleus, S_d:
remainder of eardrum; (b) Rigid
pistons with mechanical
coupling; (c) Network with
ideal transformer.*

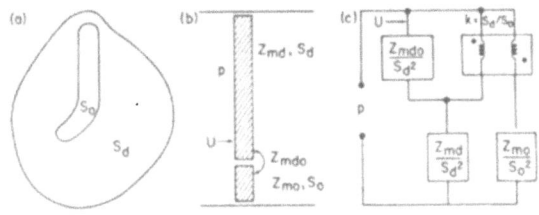

impedance. To proceed to higher frequencies, it is essential to accommodate
the mechanical complexity of the eardrum revealed in the holographic studies
of Tonndorf and Khanna (1972), Løkberg et al (1980) and others.

Fig. 5 shows a lumped element representation of the human eardrum which takes
into account some of its major vibrational characteristics while avoiding the
fine structure associated with the flexural resonances which must surely be
present. First, the eardrum surface is divided into two zones: S_o, which is
tightly coupled to the malleus at all frequencies, and S_d, which is free to
move with a substantial measure of independence at high frequencies where its
motion is controlled primarily by its own inertia. These zones are treated
as rigid pistons with mechanical impedances Z_{md} and Z_{mo} and they are
mechanically coupled by an impedance Z_{mdo}. (At low frequencies, this
coupling impedance represents the stiffness between the central and outer
portions of an elastic shell. At high frequencies, it seems to be determined
primarily by the internal damping of the shell.) Since the pistons are also
acoustically coupled through the ear canal and middle ear cavities, the
network representing the system must include an ideal transformer as shown in
Fig. 5(c).

In the complete middle ear network presented in Fig. 6, the acoustical
impedances $Z_d = Z_{md}/S_d^2$, $Z_o = Z_{mo}/S_o^2$ and $Z_{do} = Z_{mdo}/S_d^2$ are now expressed as
individual circuit elements. The elements connected to the secondary side of
the transformer (R_o, L_o etc.) correspond closely to the elements in
Zwislocki's network though the numerical values are very different due to the
differences in reference areas. On the primary side, however, L_d now
represents the inertance (mass/area²) of the major portion S_d of the eardrum,
C_d the acoustical capacitance associated with the periphery of the eardrum
and R_d the peripheral damping. The acoustical elements C_{do} and R_{do},
representing mechanical coupling between the two areas of the eardrum, are
discussed later.

With suitable choices of parameters (e.g. Shaw and Stinson 1981),
calculations based on Fig. 6 are in reasonable agreement with the median

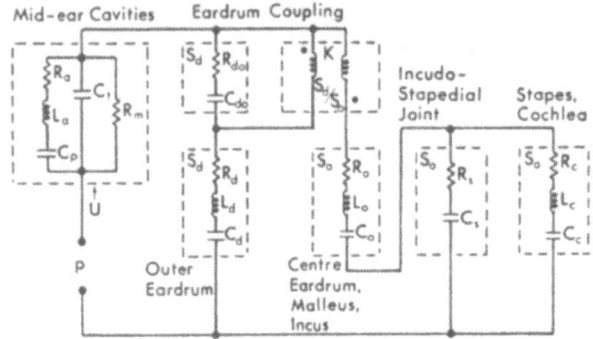

Fig. 6. *Middle ear network based on compound-eardrum concept.*

input impedance data for normal and pathological ears, recent estimates of ear canal standing wave ratio (Stinson et al 1982) and observations of eardrum vibration.

Some further implications of this network are presented in Fig. 7. The seven zones indicate the fractions of incident energy that are absorbed in various parts of the ear and the fraction that is reflected at the eardrum when a progressive wave of unit energy enters the ear canal. The fraction absorbed by the cochlea rises to a maximum of about 26% at 1 kHz while the fraction reflected at the eardrum falls to a minimum of approximately 28% at 4 kHz. At frequencies greater than 2.5 kHz, the lion's share of the energy is taken by R_{do} which, from its position in the network, must surely represent mechanical resistance within the eardrum. This conclusion is perhaps surprising but seems inescapable if one accepts the essential correctness of the standing wave data.

The probable function of R_{do} is revealed in Fig. 8 which shows the piston velocity ratio as a function of frequency for various values of this

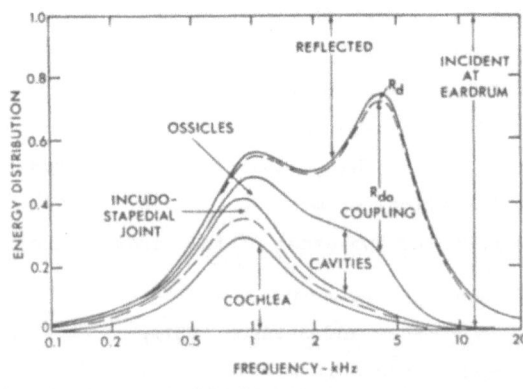

Fig. 7. *Calculated fractions of incident energy absorbed by various structures and fraction reflected at eardrum. Based on middle-ear network with values given in Shaw and Stinson (1981).*

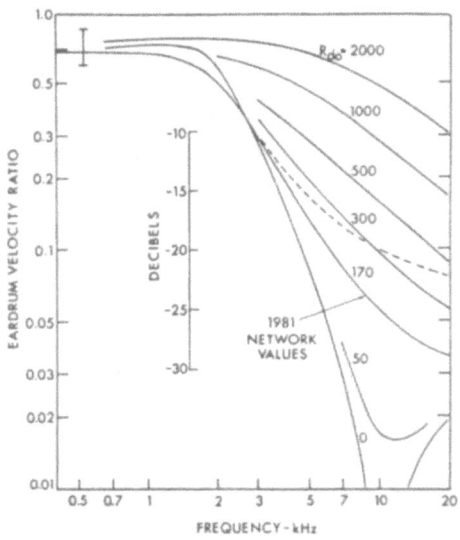

Fig. 8. *Calculated eardrum velocity ratio (Velocity of area S_o/Velocity of area S_d) - for various values of R_{do}. All other network parameter values as given in Shaw and Stinson (1981). Broken line based on R_{do} varying as square root of frequency.*

·resistance. At frequencies greater than 10 kHz this ratio (hence, the stapes velocity ratio also) is nearly proportional to R_{do}. Evidently, at high frequencies, middle ear transmission is enhanced by the presence of mechanical damping in the eardrum. When R_{do} is 170 ohms, as in Fig. 7, the piston velocity ratio is approximately 0.7 at frequencies up to 2 kHz, in agreement with Tonndorf and Khanna, and then falls rapidly with increasing frequency passing through the value of 0.14 at 5 kHz which also appears to be in agreement with experiment.

When one considers the viscoelastic nature of the eardrum, it would not be surprising if R_{do} and related elements such as C_{do} should prove to be frequency dependent. The broken line in Fig. 8 has been drawn on the assumption that C_{do} is constant while R_{do} increases with the square root of frequency. The result is a significant increase in the velocity ratio at high frequencies.

5. EARDRUM ASYMMETRY, HEARING THRESHOLD LEVELS AND THERMAL NOISE

The smoothness of the median free-field hearing threshold curve suggests the presence of a mechanism, in the middle ear or beyond, which counterbalances the principal resonance of the human ear (Shaw, 1982). Work now in progress, inspired by the evident lack of symmetry between the anterior and posterior zones of the eardrum, starts with the premise that the major area, identified as S_d in Fig. 5(a), should be divided. This leads to a three-piston model which, with a suitable choice of parameters, produces the required minimum in

malleus velocity at 2.7 kHz. It is, however, necessary to accept less
eardrum damping than seems likely in view of the standing wave data.

In the free field, the radiation impedance of the external ear shunts the
input terminals of the middle ear network. When the thermal noise of the
combined system is calculated by invoking the Nyquist noise generator
theorem, it is found that most of the noise appearing at the oval window is
associated with the cochlea. By estimating the transmission of sound energy
from the eardrum to the cochlea (e.g. lower solid curve in Fig. 1), it is
then possible to calculate the detection limit imposed by thermal noise given
some knowledge of the signal detectability (e.g. Green et al 1959). At 500
Hz this limit is approximately 20 dB below the observed median hearing
threshold level at the eardrum. Between 8 and 16 kHz, however, the thermal
curve matches recently-determined threshold levels for young ears (Shaw and
Stinson 1980) which is surprising.

REFERENCES

Green, D.M., McVey, M.J., and Licklider, J.C.R. (1959). Detection of a
    pulsed sinusoid in noise as a function of frequency, *J. Acoust. Soc.
    Am.* 31, 1446-1452.
Løkberg, O.J., Høgmoen, K. and Gundersen, T. (1980). Vibration
    measurement in the human tympanic membrane - in vivo, *Acta Otolaryngol.*
    89, 37-42.
Shaw, E.A.G. (1979). Performance of external ear as sound collector, *J.
    Acoust. Soc. Am. Suppl. 1*, 65, S9.
Shaw, E.A.G. (1980). The Acoustics of the External Ear. In: *Acoustical
    Factors Affecting Hearing Aid Performance*, edited by G.A. Studebaker
    and I. Hochberg (University park Press, Baltimore), pp. 109-125.
Shaw, E.A.G. (1982). The 1979 Rayleigh Medal Lecture: The Elusive
    Connection. In: *Localization of Sound: Theory and Applications*,
    edited by R.W. Gatehouse (Amphora Press, Groton, Conn.), pp. 13-29.
Shaw, E.A.G. and Stinson, M.R. (1980). Middle-ear function, thermal noise
    and hearing threshold levels. *Proceedings of the Tenth International
    Congress on Acoustics* (Australian Acoust. Soc., Sydney) Vol.2, p.B-3.4.
Shaw, E.A.G. and Stinson, M.R. (1981). Network concepts and energy flow in
    the human middle-ear, *J. Acoust. Soc. Am. Suppl. 1* 69, S43.
Stinson, M.R. and Shaw, E.A.G. (1982). Wave effects and pressure
    distribution in the ear canal near the tympanic membrane. *J. Acoust.
    Soc. Am. Suppl. 1*, 71, S88.
Stinson, M.R., Shaw, E.A.G. and Lawton, B.W. (1982). Estimation of
    acoustical energy reflectance at the eardrum from measurements of
    pressure distribution in the human ear canal, *J. Acoust. Soc. Am.* 72,
    766-773.
Tonndorf, J. and Khanna, S.M. (1972). Tympanic membrane vibrations in human
    cadaver ears studied by time-averaged holography, *J. Acoust. Soc. Am.*
    52, 1221-1233.
Zwislocki, J. (1962). Analysis of the Middle-Ear Function. Part I: Input
    Impedance, *J. Acoust. Soc. Am.* 34, 1514-1523.

MODELING REVERSE MIDDLE EAR TRANSMISSION OF ACOUSTIC DISTORTION SIGNALS

John W. Matthews

Computer Systems Laboratory
Washington University
724 S. Euclid Ave.
St. Louis, Missouri 63110, U.S.A.

ABSTRACT

In modeling the propagation of signals produced in the cochlea, the effects of the middle ear must be included. We present a linear two-port network model of the middle ear of cat with air cavities open. Effects of the eardrum, ossicular chain, oval and round windows, and fluid in the vestibule are included. The two ports represented are: 1) the ear canal; and 2) the basal end of the cochlear spiral. The model parameters were selected to fit experimental data measuring various aspects of forward transmission only. However, we have used the model to reproduce acoustic distortion signals observed in the ear canal. Three significant findings are that: 1) The design of an acoustic coupler can have a large effect on signals measured in the ear canal; 2) The middle ear and acoustic coupler affect the reflection of distortion signals back into the cochlea and therefore affect distortion signals observed within the cochlea as well as in the ear canal; 3) The reverse transmission properties of the middle ear circuit model are highly frequency dependent.

1. INTRODUCTION

Kemp's (1978) observation of acoustic emissions from the ear indicates that sound can propagate "backwards" through the middle ear into the ear canal. This paper deals with modeling the effects of the middle ear on the propagation of distortion products generated within the cochlea. This middle ear modeling is part of a larger modeling effort (Matthews, 1980; Matthews et al., 1981) whose objective is the consistent interpretation of intracochlear and aural acoustic distortion signals observed in response to steady-state two-tone stimuli (Kim et al., 1980; Siegel et al., 1982; Kim, 1980). These intermodulation distortion signals (e.g. $2f_1-f_2$) are interpreted as being generated in the cochlea at a locus where both primary frequencies (f_1 and f_2) have strong response; the distortion signals then propagate within the cochlea both apically to the characteristic place of the distortion signal and basally to the stapes, through the middle ear, and into the ear canal.

2. THE COMPREHENSIVE MODEL

This section presents a "comprehensive" model which includes the effects of the stimulus delivery system, the middle ear, and the mechanics of the cochlea for cat (see Fig. 1). Since in this comprehensive model only the cochlea is nonlinear, any distortion signals must be generated within the cochlea. Distortion signals appearing at the eardrum must be propagated out of the

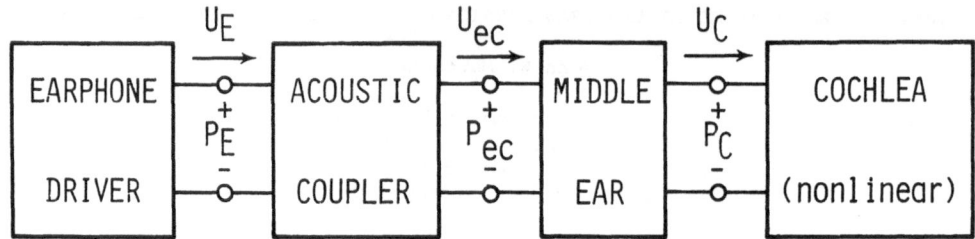

Fig. 1. Block diagram of a "comprehensive" model of the peripheral auditory system of cat and a stimulus delivery system. Pressure (P) and volume velocity (U) are indicated at the earphone driver (E), ear canal near the eardrum (ec), and the most basal position of the cochlear spiral (C). Note that in this model only the cochlea is nonlinear.

cochlea through the middle ear. On the other hand, the acoustic impedance of the driver, coupler and middle ear (as seen looking out of the cochlea) will determine the amount of reflection of these distortion signals back into the cochlea and thus will affect the total distortion signal within the cochlea.

The models for the driver, coupler and middle ear will be electrical circuits and an acoustic-electric analogy will be maintained throughout the paper; this analogy is summarized in Table 1. These acoustic variables and units will be used throughout this paper, even in the middle ear model where mechanical variables (e.g., force and velocity) might be more natural.

The driver and acoustic coupler models are those developed by Matthews (1980) to represent the physical devices used by Kim et al. (1980). The models are equivalent to a series RLC circuit connected to the "input" port of the middle ear model as far as their effect on signals produced in the cochlea is concerned. Element values for this RLC circuit are given in Fig. 4.

Within the cochlear model, the fluid is represented by a two-dimensional, linear, ideal fluid. The basilar membrane is represented by mass, stiffness, and damping functions of distance along the cochlear spiral. The damping of the basilar membrane is also a function of the velocity of the basilar membrane and this causes the cochlear model to be nonlinear (see Matthews, 1980).

Variable Type	Units	Variable Type	Units
pressure	$dyne/cm^2$	voltage	volts
volume velocity	cm^3/sec	current	amps
acoustic compliance	$cm^5/dyne$	capacitance	farads
acoustic mass	gm/cm^4	inductance	henries
acoustic damping	$dyne\text{-}sec/cm^5$	resistance	ohms
acoustic impedance	$dyne\text{-}sec/cm^5$	impedance	ohms

Table 1. Acoustical-Electrical analogy and units. These units are used throughout the paper.

3. THE MIDDLE EAR MODEL

Figure 2 shows a circuit model for the middle ear of cat with bulla and tympanic air cavities widely open. This model modifies and extends the circuit

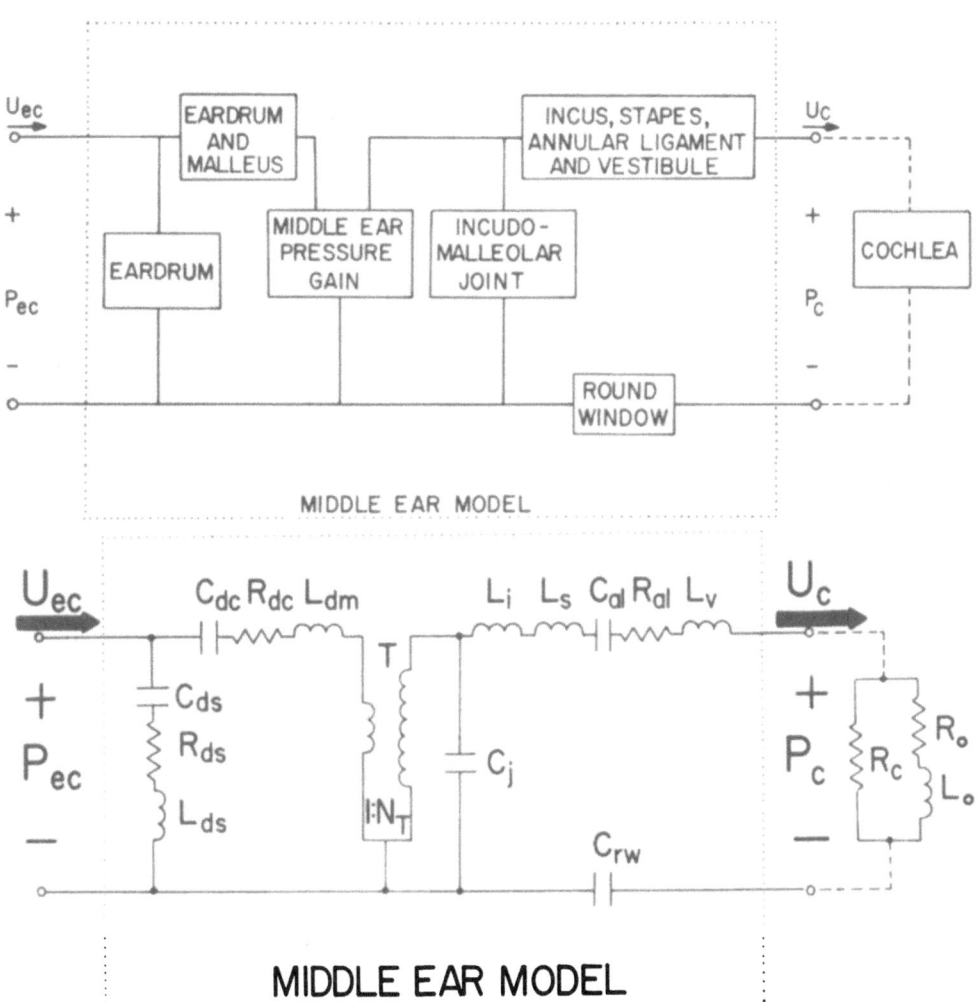

Fig. 2. Circuit model for the middle ear of cat with air cavities open.
Upper: block diagram; Lower: circuit. P_{ec} is the pressure across the eardrum;
U_{ec} is the volume velocity into the eardrum. P_C is the pressure across the
cochlear partition at the most basal position of the cochlear spiral; U_C is
the volume velocity into scala vestibuli at this same position. R_C, R_o and L_o
represent a model for the input impedance of cat cochlea and are not part of
the middle ear model. The portion of the circuit to the right of L_i is after
Lynch et al. (1982). The values of the elements: $C_{ds} = 8 \times 10^{-8}$; $R_{ds} = 1300$;
$L_{ds} = 0.054$; $C_{dc} = 3.5 \times 10^{-7}$; $R_{dc} = 55.2$; $L_{dm} = 0.04$; $N_T = 55$; $C_j = 1.2 \times 10^{-11}$; $L_i = 1.6$; $L_s = 3.3$; $L_v = 22$; $C_{al} = 3.7 \times 10^{-10}$; $R_{al} = 2 \times 10^5$; $R_C = 1.2 \times 10^6$; $R_o = 2.8 \times 10^5$; $L_o = 2250$.

14

presented by Peake and Guinan (1967). The parameters C_{ds}, R_{ds}, and L_{ds} represent the compliance, damping and inertial effects of any motion of the eardrum that is not coupled to the malleus. C_{dc}, R_{dc}, and L_{dm} represent the compliance, damping and inertial effects of the coupled eardrum-malleus motion. T represents the pressure gain of the middle ear due to the lever action of malleus-incus motion and the difference in the effective areas of the eardrum

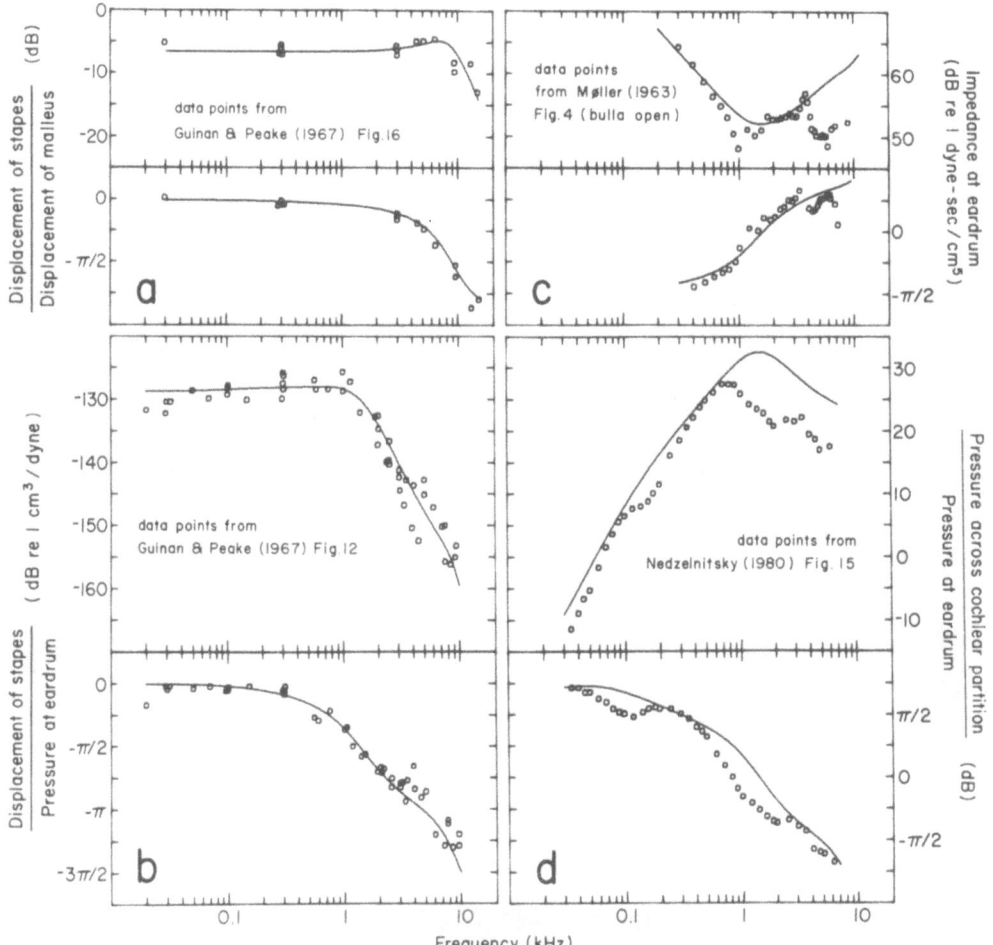

Fig. 3. *Comparison of experimental data and response of the middle ear model. The data points are experimental data for cat plotted versus frequency (from various authors as cited in the figure). Magnitude in dB (upper) and phase in radians (lower) is shown in each part. Part (a): ratio of linear stapes displacement to linear malleus displacement; part (b): ratio of linear stapes displacement to pressure across the eardrum; part (c): acoustic input impedance at the eardrum; part (d): ratio of the pressure across the base of the cochlear partition to the pressure across the eardrum. The data in parts (a) to (c) were used in selecting element values for the model, but the data in part (d) were not used.*

and the stapes footplate. C_j represents compliance in the incudo-malleolar joint. L_i, L_s, and L_v represent inertial effects of the incus, stapes and fluid in the vestibule of the inner ear, respectively. C_{al} and R_{al} represent compliance and damping of the of the annular ligament around the footplate of the stapes. C_{rw} represents compliance effects of the round window. R_c, R_o, and L_o represent a model of the input impedance of the cochlea of cat from Lynch et al. (1982). A mechanical middle ear model, equivalent to the circuit model shown here, is presented by Neely (1981).

Figure 3 compares the response of the circuit model shown in Fig. 2 to various experimental data. The data in Fig. 3a, b, and c, but not those in Fig. 3d, were used in determining element values using methods similar to those used by Peake and Guinan (1967). Therefore, Fig. 3d shows an independent test of the middle ear model.

All of the data in Fig. 3 are measures of forward transmission through the middle ear terminated by the cochlea. Figure 4 shows the response of the middle ear circuit when driven "backwards" and terminated with a series RLC circuit representing the driver and coupler models. Figure 4a shows the magnitude (dB re 1 acoustic ohm) and phase of the acoustic impedance looking out of the cochlea. For comparison the impedance looking into the cochlea for frequencies between 600 and 6000 Hz is approximately constant with a value of about 122 dB and zero phase (see Lynch et al., 1982; Matthews, 1980). Figure 4b shows the "reverse pressure gain" of the middle ear circuit model. Both panels show the circuit response for three different ear canal terminations.

4. ILLUSTRATION OF THE USE OF THE COMPREHENSIVE MODEL

Figure 5 illustrates an application of the comprehensive model. A volume velocity containing only two frequencies, f_1 and f_2, varied together such that $(2f_1-f_2) = 1550$ Hz, was supplied by the earphone driver. The magnitudes and phases of f_1 and f_2 were selected such that their components in P_{ec} were both 65 dB SPL, zero cosine phase. The 1550 Hz components of both P_{ec} and the basilar membrane displacement at the 1550 Hz place were computed for the model. The basilar membrane displacement is plotted as the equivalent value of P_{ec} for a 1550 Hz single tone necessary to produce the same displacement. The ear canal pressure results of the model show an overall similarity to Kim's (1980, Fig. 10) experimental data from cat, with an exception that the cat data show a prominent "lobing" while the model results do not. The basilar membrane displacement results of the model appear consistent with human psychophysical results by Zwicker (1980, Fig. 1 and 2).

5. DISCUSSION

For the results shown in Fig. 5, the middle ear circuit accurately models both
forward and reverse transmission. However, there are weaknesses in the middle
ear model. Most notable is the eardrum "shunt" represented by R_{ds}, C_{ds} and
L_{ds}. The acoustic impedance looking into the eardrum changes abruptly between
3 and 4 kHz which is not reproduced by the model (see Fig. 3c). A more elabo-
rate model for eardrum coupling (e.g. Shaw and Stinson, 1981) might improve
the model in this regard.

Distortion signals computed for the model's "ear canal" are particularly sensitive to both the form and parameter values of the coupler, driver, and

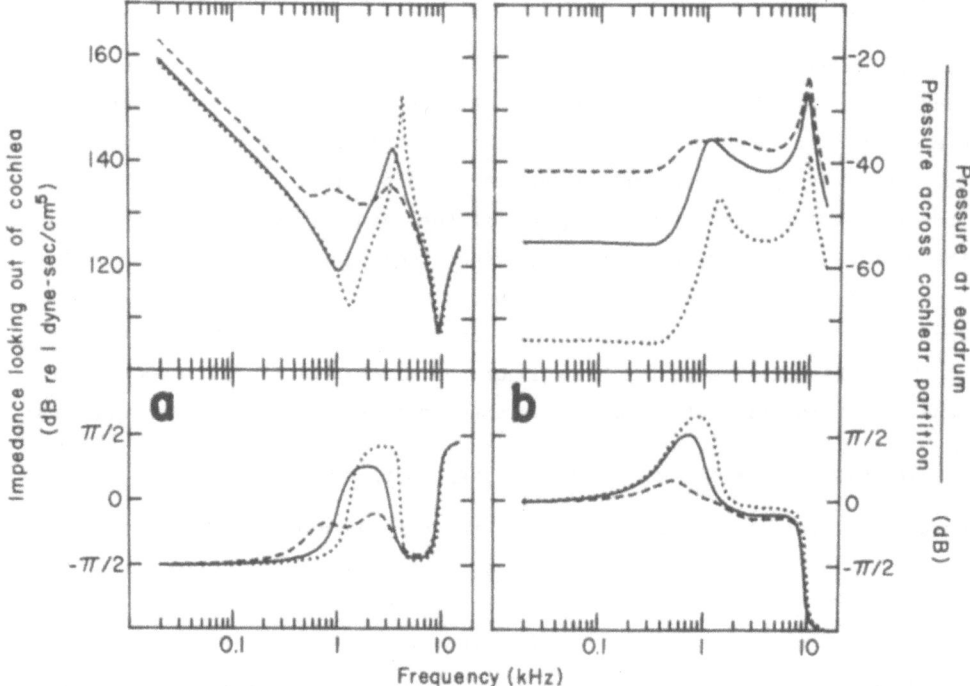

Fig. 4. *Reverse response of the middle ear model. Two types of response of
the middle ear circuit model are shown for a source at the cochlear port and
three different passive terminations at the ear canal port. Part (a): the
acoustic impedance looking out of the cochlea; part (b): the reverse pressure
gain of the middle ear model. Magnitude in dB (upper) and phase in radians
(lower) are shown. The solid lines are when the ear canal port of the circuit
is terminated with the earphone driver and acoustic coupler models used by
Matthews (1980). This is equivalent to a series RLC circuit with values: R =
140; L = 0.0434; and C = 2.28 x 10⁻⁶. The other curves show the effect of
changing the driver/coupler model to have 10 times (dashed) or 0.1 times (dotted) the impedance of the Matthews model.*

eardrum models. (This effect is shown for coupler-driver models in Fig 4b). Hence, different stimulus delivery systems could show very different distortion signals in the ear canal of the same animal because they "load" the auditory system differently. Even distortion propagation within the cochlea can be affected by the middle ear and stimulus delivery system through their effect on cochlear "loading" (Fig. 4a). Therefore, it is important that attention be given to the middle ear and stimulus delivery system when studying signals produced in the cochlea.

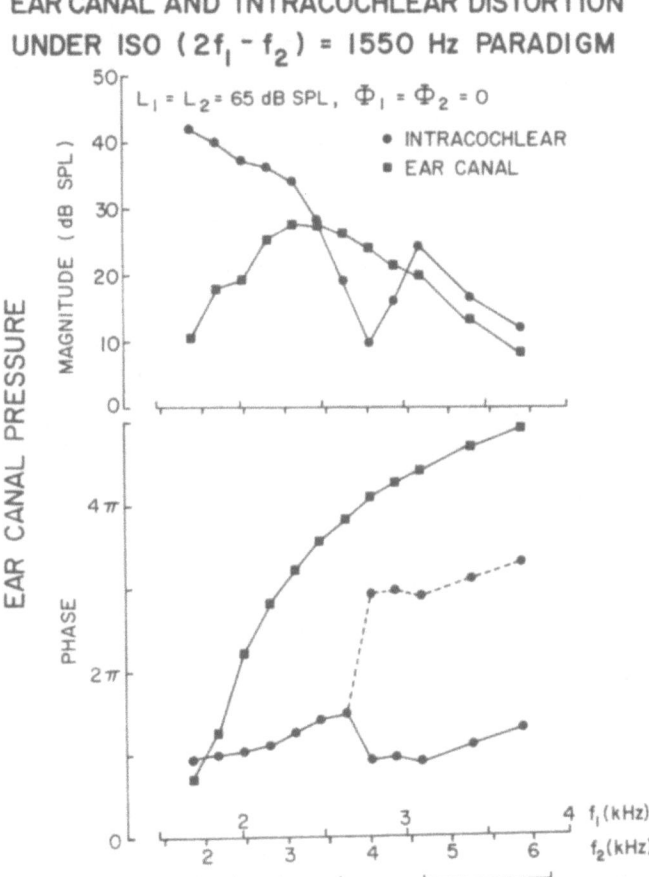

Fig. 5. Generation and propagation of distortion signals in the comprehensive model. The squares show the magnitude and phase of the 1550 Hz component of the ear canal pressure when the stimulus contained only f_1 and f_2. The circles show the magnitude and phase of a measure of the 1550 Hz component of the cochlear partition displacement at the characteristic place for 1550 Hz. The measure plotted is the ear canal pressure of a single 1550 Hz tone necessary to produce the same displacement at the characteristic place.

18

Acknowledgments

Some of this work was contained in the author's doctoral dissertation and was presented at the 101st meeting of the Acoustical Society of America. The Author thanks D.O. Kim, E.L. LePage, C. E. Molnar, and S.T. Neely for helpful comments and D. Bozzay for assistance in figure preparation. This work was supported under NIH grants NS-07498, RR-00396, and RR-01379.

REFERENCES

Guinan, J.J.,Jr. and Peake, W.T. (1967). Middle-Ear Characteristics of Anesthetized Cats, *The Journal of the Acoustical Society of America*, 41, 1237-1261.

Kemp, D.T. (1978). Stimulated acoustic emissions from within the human auditory system, *The Journal of the Acoustical Society of America*, 64, 1386-1391.

Kim, D.O. (1980). Cochlear mechanics: Implications of electrophysiological and acoustic observations, *Hearing Research*, 2, 297-317.

Kim, D.O., Molnar, C.E., and Matthews, J.W. (1980). Cochlear mechanics: Nonlinear behavior in two-tone responses as reflected in cochlear-nerve-fiber responses and in ear-canal sound pressure, *The Journal of the Acoustical Society of America*, 67, 1704-1721.

Lynch, T.J.,III, Nedzelnitsky, V., and Peake, W.T. (1982). Input impedance of the cochlea of cat, *The Journal of the Acoustical Society of America*, 72, 108-130.

Matthews, J.W. (1980). *Mechanical Modeling of Nonlinear Phenomena Observed in the Peripheral Auditory System*. Doctoral dissertation, (Washington University, St. Louis, Missouri).

Matthews, J.W., Kim, D.O., Molnar, C.E., and Neely, S.T. (1981). Modeling reverse middle ear transmission: Aural acoustic distortion products, "echoes," and spontaneous emissions, *The Journal of the Acoustical Society of America*, 69, S43, (A).

Møller, A.R. (1963). Transfer Function of the Middle Ear, *The Journal of the Acoustical Society of America*, 35, 1526-1534.

Nedzelnitsky, V. (1980). Sound pressure in the basal turn of the cat cochlea, *The Journal of the Acoustical Society of America*, 6, 1676-1689.

Neely, S.T. (1981). *Fourth-order Partition Mechanics for a Two-dimensional Cochlear Model*. Doctoral dissertation, (Washington University, St. Louis, Missouri).

Peake, W.T. and Guinan, J.J.,Jr. (1967). Circuit Model for the Cat's Middle Ear, *Massachusetts Institute of Technology Research Laboratory of Electronics Quarterly Progress Report*, 84, 320-326.

Shaw, E.A.G. and Stinson, M.R. (1981). Network concepts and energy flow in the human middle-ear, *The Journal of the Acoustical Society of America*, 69, S43, (A).

Siegel, J.H., Kim, D.O., and Molnar, C.E. (1982). Effects of altering organ of Corti on cochlear distortion products f_2-f_1 and $2f_1-f_2$, *Journal of Neurophysiology*, 47, 303-328.

Zwicker, E. (1980). Cubic difference tone level and phase dependance on frequency difference and level of primaries. In: *Psychophysical, Physiological and Behavioural Studies in Hearing*, edited by G. van den Brink and F.A. Bilsen, (Delft University Press, Delft, The Netherlands), pp. 268-271.

RECENT DEVELOPMENTS IN MODELLING THE EARDRUM AND RELATED STRUCTURES USING THE FINITE-ELEMENT METHOD

W. Robert J. Funnell

BioMedical Engineering Unit, McGill University
3655 Drummond Street, Montréal, Québec, Canada H3G 1Y6

ABSTRACT

The finite-element method allows one to model a structure as an assemblage of simple elements, using a digital computer. Its strong point is its ability to handle complexities, nonuniformities and irregularities such as abound in living systems. This paper discusses some recent developments in the use of this method to model the eardrum and related structures, including the generation of eardrum models with various mesh resolutions; their use in studying the system's natural frequencies and the effects of damping; and the creation of models of auditory structures other than the eardrum.

1. GENERATION OF EARDRUM MODELS

The mathematical models being discussed here are based on the 'finite-element method'. This is a method of analysis which has been used in engineering for a number of years. More recently it has been applied to biological problems. It is well suited to such problems because its strong point is its ability to handle complexities, nonuniformities and irregularities such as abound in living systems. The method handles a complicated system by dividing it into a large number of simple parts. Each part can be analyzed relatively easily, and its characteristics can be expressed with a small matrix equation. The interactions among the parts, and thus the over-all behaviour of the system, can then be calculated by assembling the small matrices into one large matrix equation suitable for solution by computer.

My finite-element calculations are done using a modified version of SAP on the PDP-11/70 time-sharing system (MEDNET) of the BioMedical Engineering Unit at McGill University. SAP is a powerful general-purpose finite-element programme that was developed at the University of California (Bathe, Wilson and Peterson, 1974) for large CDC and IBM computers.

Until recently all of my finite-element meshes for the eardrum were generated manually. This made the generation of new meshes extremely tedious and prone to error, and made it impractical to attempt a study of the effects of varying the fineness of the mesh. The automated mesh-generation schemes usually used are not well suited to irregularly shaped biological structures, so I have developed a hierarchical modelling scheme for three-dimensional shallow shells with variable mesh resolution. This scheme uses automatic two-dimensional finite-element mesh generation, and a specialized method for implementing the curvature of the eardrum (Funnell, 1983b).

Using the above system I have examined the convergence of the drum model as the mesh resolution is increased. Using as a measure of resolution the nominal number of elements across the diameter of the structure, I varied the resolution from 6 to 15, my previous manually-generated model (Funnell and Laszlo, 1978) having been equivalent to about 8. Figure 1 shows some of the automatically generated meshes. I concluded that although the results at 8 were of a precision comparable to other sources of error in the model, a resolution of 12 is a better compromise between precision and complexity (Funnell, 1983b).

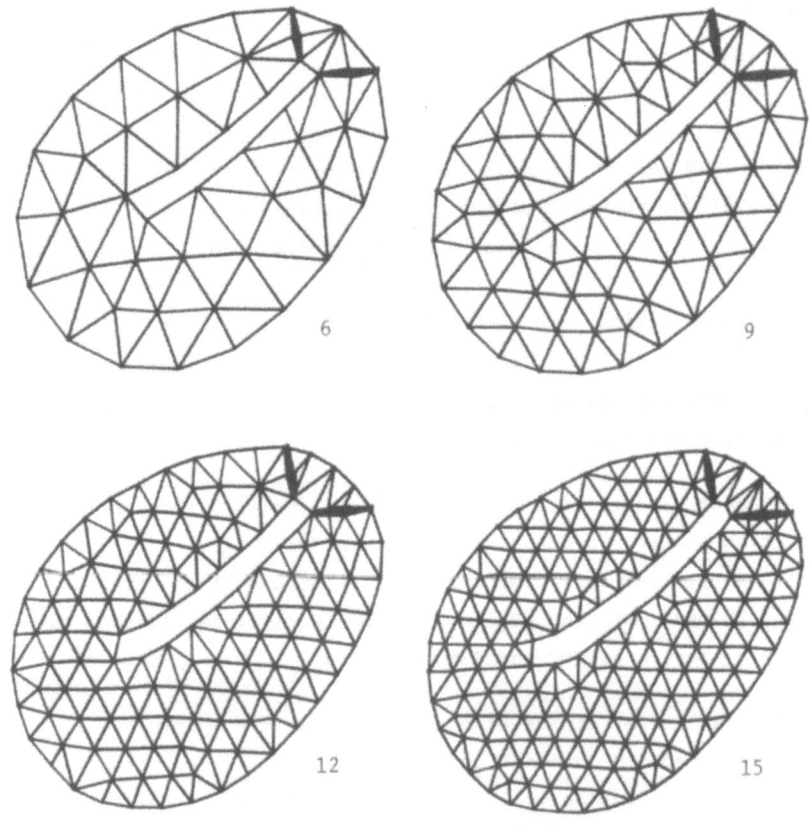

Fig. 1. Eardrum meshes generated for resolution parameters of 6, 9, 12 and 15. See text for definition of resolution parameter.

2. NATURAL FREQUENCIES

Using the models generated with the above method, an earlier study on the effects of parameter variations on the undamped natural frequencies and modes of the eardrum model (Funnell, 1980) was redone and extended (Funnell, 1983b). The parameters examined were (1) the angular stiffness and (2) the moment of inertia representing the ossicular load; (3) the material stiffness, (4) the mass density, and (5) the thickness of the eardrum itself; and two parameters representing the shape of the eardrum -- (6) the depth and (7) the curvature. The effects of varying the first five parameters were relatively straightforward and qualitatively predictable: increasing stiffnesses raised the natural frequencies, increasing inertias lowered them, and the effects of the ossicular parameters were much less than those of the eardrum parameters. The effects of varying the three-dimensional shape were more complex, and seem to indicate that both the conical nature and the curvature of the drum may serve to broaden its frequency response (Fig. 2).

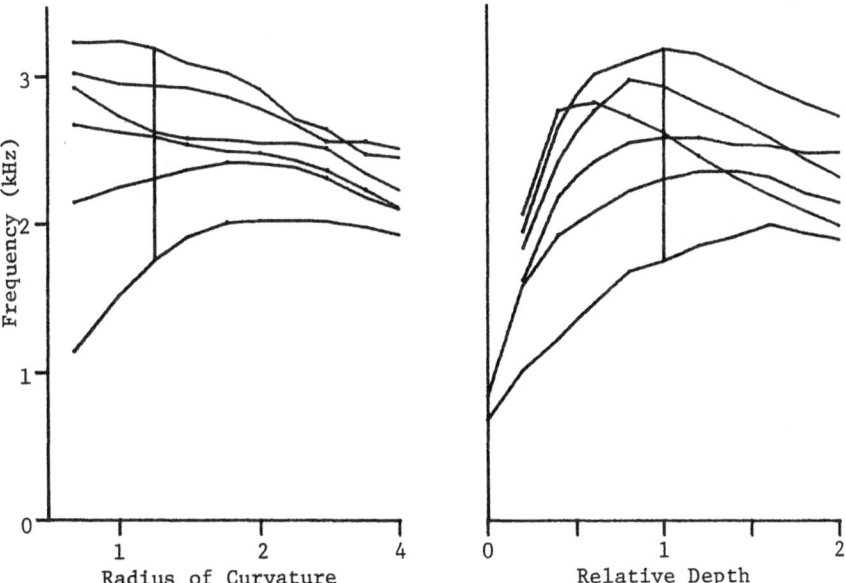

Fig. 2. Lowest six natural frequencies of eardrum model, as functions of (a) radius of curvature and (b) relative depth. The fact that some of the lines cross indicates that the relative order of the different modes may change. The vertical lines indicate the 'normal' values for the curvature and depth.

3. DAMPING

The natural frequencies discussed above were calculated in the absence of any damping. In order to be able to consider the amplitude of the eardrum's response to arbitrary frequencies, rather just looking at its natural frequencies and mode shapes, it is necessary to include viscous effects, or damping, in the model.

The SAP programme offers two alternative approaches to damping: either superposition of the undamped natural modes using one extra parameter (fraction of critical damping, taken to be the same for every mode); or direct time-domain integration using two damping parameters (Rayleigh damping). The superposition method is computationally cheaper if a reasonably small number of modes is adequate to represent the system response (Nelson and Greif, 1975, for example) but the fact that the natural frequencies of the eardrum model are quite closely spaced means that a fairly large number of modes must be included.

I have done some preliminary calculations with the direct-integration approach, using the same eardrum mesh as in the natural-frequency calculations. The effective damping matrix [c] is given in terms of the mass and stiffness matrices as

$$[c] = a_0[m] + a_1[k],$$

where a_0 and a_1 are the two independent damping parameters that may be specified. It can be shown (Nelson and Greif, 1975) that the resultant damping ratio for the i-th mode, with angular frequency w_i, is given by

$$b_i = (a_0/w_i + a_1 w_i)/2.$$

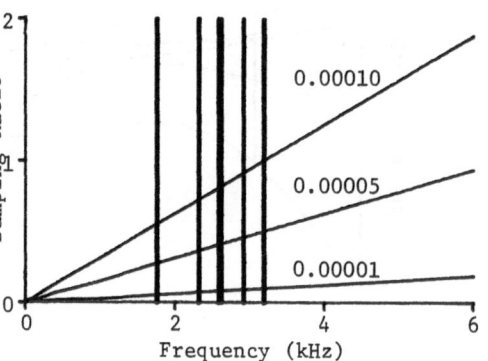

Fig. 3. *Damping ratio as a function of frequency for the three values of a_1 used here. The vertical bars indicate the values of the first six natural frequencies. Note that the third and fourth bars overlap.*

In the calculations discussed below I have set $a_0 = 0$, and used values for a_1 of 10, 50 and 100 times 10^{-6}. Figure 3 shows the damping ratio as a function of frequency for these values of the damping parameters. Also noted are the frequencies of the first six undamped modes of the eardrum model.

I have simulated the effect of applying a step function of torque to the rotational axis of the manubrium, and calculated the time response of the manubrium itself and of several points on the surface of the drum. The time responses were calculated in steps of 0.02 ms from 0 to 7 ms. Using NEXUS, a general-purpose systems-and-signal-analysis package (Kearney and Hunter, 1982), I differentiated these step responses to obtain the impulse responses, and then computed the Fourier transforms.

Figure 4 shows the resulting frequency responses for one particular point on the eardrum for the different values of a_1. It can be seen clearly that the multiple undamped modes of the system are smeared out by the damping. Note that these results are preliminary only: the computations have not been rigourously checked. The overall frequency response is also affected by the finite-element mesh and by the step-by-step integration procedure.

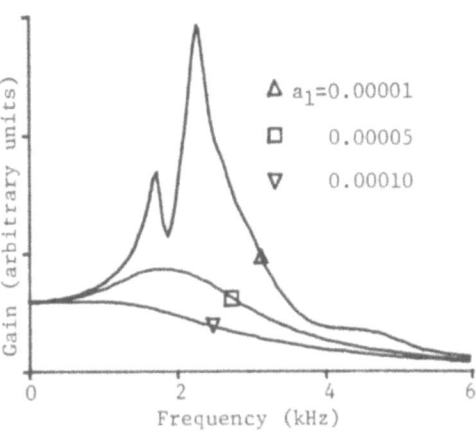

Fig. 4. Frequency responses of a particular point on the eardrum, for three values of the damping parameter a_1.

It may become necessary to implement damping representations different from those currently used in SAP, and ultimately to use a more general frequency-dependent complex-modulus approach (Gupta, 1974; Soni and Bogner, 1982). Although practically nothing is known experimentally about material damping coefficients in the eardrum except for some low-frequency estimates by Decraemer (see Funnell and Laszlo, 1982), it will be useful to be able to estimate the effects of levels of damping that at least are consistent with what is known about viscoelastic behaviour in collagenous tissues. Middle-ear models with simplified lumped eardrum representations suggest that the damping in the eardrum is quite large (Shaw and Stinson, 1981), and there may be some methodological difficulties stemming from the fact that most engineering structural analyses concentrate on relatively low levels of damping, treated as perturbations of the undamped case.

4. MODELS OF OTHER STRUCTURES

We have in the past presented a finite-element model of the middle-ear ossicles, constructed on the basis of serial histological sections (Funnell and Phelan, 1981). More recently a student, V. Goel, has constructed simple models of the cat posterior incudal ligament and annular ligament.

Figure 5 shows the model of the incudal ligament. Preliminary simulations with SAP produced an estimate of 2300 dyn cm for the effective angular stiffness of the incudal ligament, to be compared to my previous estimate, based on a much more simplified geometry (Funnell and Laszlo, 1978) of 8500 dyn cm.

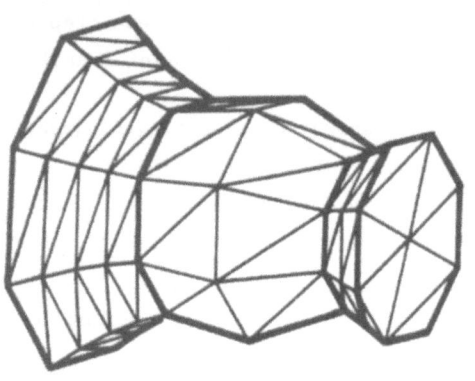

This modelling will be pursued using a more complete series of histological sections: contours from this material will be digitized and used to create a much more accurate finite-element representation of the ossicular ligaments and muscles than has been possible before.

Fig. 5. Finite-element model of the posterior incudal ligament. The three parts shown are, from left to right, the lateral portion of the ligament, the incus, and the medial portion of the ligament.

The generation of element meshes for data from serial sections is extremely tedious if done manually. Several papers have been published recently describing approaches to three-dimensional mesh generation (Nguyen, 1982; and Perucchio, Ingraffea and Abel, 1982, among others), but none are very well suited to highly irregular three-dimensional objects. A number of methods have been described for triangulating irregular three-dimensional <u>surfaces</u> defined by contours from serial sections (Funnell, 1983a), but for modelling of the mechanics of the structures one must generate internal meshes for three-dimensional <u>solids</u>. It is necessary to extend methods like that discussed in Section 1 above, which generate triangular elements inside two-dimensional areas, to the generation of tetrahedral elements inside three-dimensional volumes.

Acknowledgements

This work has been supported by the Medical Research Council of Canada.

REFERENCES

Bathe, K.-J., Wilson, E.L., and Peterson, F.E. (1974). *SAP IV. A Structural Analysis Program for static and dynamic response of linear systems.* Report No. EERC 73-11, University of California (Berkeley), vii + 59 pp. + appendices

Funnell, W.R.J., and C.A. Laszlo (1978). Modeling of the cat eardrum as a thin shell using the finite-element method. *J. Acoust. Soc. Am.* <u>63</u>, 1461 - 1467

Funnell, W.R.J. (1980). Natural frequencies of a finite-element model of the cat eardrum. *J. Acoust. Soc. Am.* <u>67</u> Suppl. 1, S88

Funnell, W.R.J., and Phelan, K.E. (1981). Finite-element modelling of the middle-ear ossicles. *J. Acoust. Soc. Am.* <u>69</u> Suppl. 1, S14

Funnell, W.R.J., and Laszlo, C.A. (1982). A critical review of experimental observations on ear-drum structure and function. *ORL* <u>44</u>, 181 - 205

Funnell, W.R.J. (1983a). On the choice of a cost function for the reconstruction of surfaces by triangulation between contours. *Comp. and Struct.*, in press

Funnell, W.R.J. (1983b). On the undamped natural frequencies and mode shapes of a finite-element model of the cat eardrum. *J. Acoust. Soc. Am.*, in press

Gupta, K.K. (1974). Eigenproblem solution of damped structural systems. *Int. J. Num. Meth. Eng.* <u>8</u>, 877 - 911

Kearney, R.E. and Hunter, I.W. (1982). NEXUS: An interpreter for the analysis of physiological systems. *Digest 9th Can. Med. and Biol. Eng. Conf.*, 47-48, Fredericton, Canada.

Nelson, F.C., and Greif, R. (1975). Damping. Pp. 603 - 623 in *Shock and Vibration Computer Programs. Reviews and summaries*, W. Pilkey and B. Pilkey (eds.), SVM-10, Shock & Vibration Information Center, U.S. Dept. Defense, viii + 663 pp

Nguyen, V.P. (1982). Automatic mesh generation with tetrahedron elements. *Int. J. Num. Meth. Eng.* <u>18</u>, 273 - 289

Perucchio, R., Ingraffea, A.R., & Abel, J.F. (1982). Interactive computer graphic preprocessing for three-dimensional finite element analysis. *Int. J. Num. Meth. Eng.* <u>18</u>, 909 - 926

Shaw, E.A.G., and Stinson, M.R. (1981). Network concepts and energy flow in the human middle-ear. *J. Acoust. Soc. Am.* <u>69</u>, S43

Soni, M.L., and Bogner, F.K. (1982). Finite element vibration analysis of damped structures. *AIAA J.* <u>20</u>, 700 - 707

Section II

Cochlear fluid mechanics

BASILAR MEMBRANE PROPERTIES AND COCHLEAR RESPONSE

C.R. Steele, J. Zais

Stanford University, Stanford CA. 94305

ABSTRACT

A WKB solution for a simplified analytical model of a guinea pig cochlea was used to investigate the effect of basilar membrane (BM) mass and orthotropy on response. The conclusion is reaffirmed that a physiological value of BM mass has little effect. The degree of orthotropy has no effect on peak location, but increases the subsequent decay of amplitude. The BM stiffness is highly dependent on the microstructure, particularly the fiber density. With such anatomical detail the localization can be estimated which agrees with direct measurement.

1. MODEL DESCRIPTION

In the 3-D box model (Fig. 1a) the BM is represented as a hinged plate in the partition which separates the two fluid-filled chambers. The "plate" simulates the BM pectinate zone in Fig. 1b, which seems to be the most significant elastic element for the gross ("first filter") response of the actual cochlea.

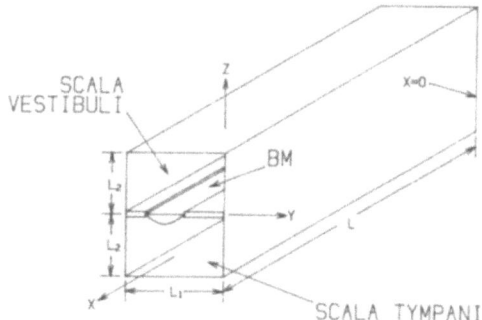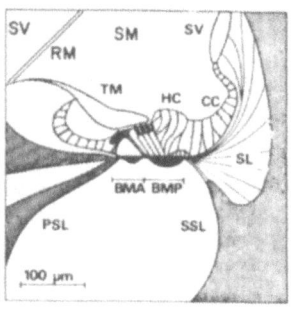

Fig. 1. a) Straightened (one-mode) model of the cochlea. The BM is modeled as a plate hinged to rigid shelves. b) Generalization of basal turn in high frequency cochleas. BMA basilar membrane arcuate zone; BMP basilar membrane pectinate zone (from Bruns (1979)).

The geometrical properties (Table 1) are adapted from Fernandez (1952). Various BM compliance distributions are shown in Fig. 2. As suggested in Steele and Taber (1981) a reasonable partition compliance is the curve C_c , which is about equal to Békésy's (1960) *post mortem* guinea pig measurements (denoted by C_B) in the apical region, and about 1/4 of those measurements in the basal region. Dancer and Franke (1980) estimate from direct *in vivo* measurements that the volume compliance in the first turn is about 1/5 that of Békésy's *post mortem* measurement. Of interest is the distribution C_{PL} , deduced from Békésy's point load tests in humans, rescaled for the guinea pig.

Table 1

X/L	b(μm)	h(μm)
0.0	60	15
0.1	97	10
0.2	115	6.8
0.3	130	4.7
0.4	140	3.8
0.5	145	3.2
0.6	150	2.9
0.7	158	2.6
0.8	165	2.2
0.9	170	1.5
1.0	175	1.0

b = BMP width
h = BMP thickness
$L_1 = L_2 = 0.8$ mm
L = 18 mm
Stapes area = 1 mm^2
d_e = viscoelastic BM damping = 0.05

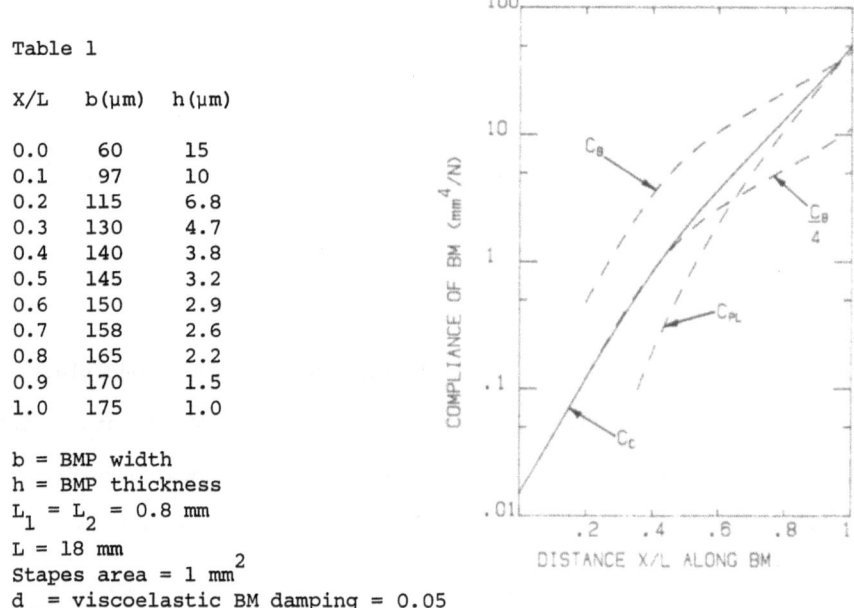

Table 1. Geometrical properties of model adapted from guinea pig measurements of Fernandez (1952).

Fig. 2. Compliance (volume displacement per unit length per unit of pressure difference) of cochlear partition: C_B , the post mortem measurement of the guinea pig by Békésy (1960); $C_B/4$, may be closer to in vivo values, especially at basal end; C_{PL} , point load tests on human cochlea (Békésy (1960)) rescaled for the guinea pig; C_C , compliance curve for Figs. 3-5.

2. SOLUTION PROCEDURE

The necessary equations for this type of model are contained in Steele and Taber (1979). The solution begins with calculating h_{eq} , the equivalent fluid thickness resisting BM motion, which depends only on the model geometry, not BM characteristics. These values of h_{eq} are then used to calculate the dispersion relation

$$f(\lambda,\omega) = \frac{\omega^2}{4} (2\rho h_{eq} + \rho_p h) - \Gamma/G = 0 \qquad (1)$$

and related quantities at 11 stations along the cochlea for a set of wavenumbers. Cubic spline coefficients are computed for these quantities, so a solution can be found for any specified frequency. All calculations were performed on an HP-85 desktop computer, demonstrating the efficiency of the WKB scheme. The approximate computing times are: calculating h_{eq}-15 minutes; solving Eq.1 and computing splines-10 minutes; solving traveling wave problem for given frequency-3 minutes.

3. RESULTS

Figure 3 shows the dispersion relation as it appears at two locations along the BM. The wave number λ is multiplied by the cross section height L_2, and the frequencies are given as a fraction of the plate cut-off frequency at the base (here about 60kHz). The three cases studied are characterized by k , the

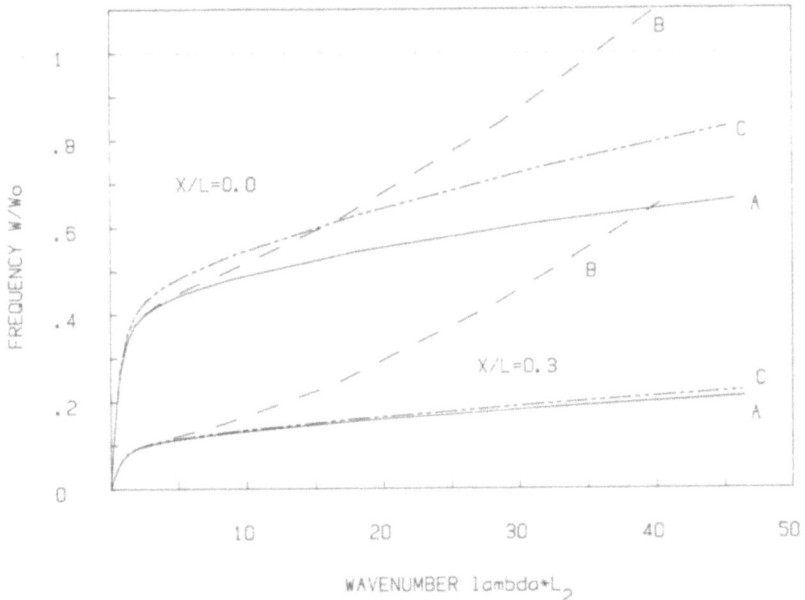

Fig. 3. Plot of dispersion relation at two locations along the cochlea for cases: A , orthotropic BM (k=0); B , isotropic BM (k=1); C , orthotropic BM with zero mass; all for compliance C_C in Fig. 2.

ratio of longitudinal to transverse stiffness: A) an orthotropic plate (k=0);
B) an isotropic plate (k=1); C) k=0 , but with a massless plate. Lighthill
(1981) argues that the dispersion plot most have continually decreasing slope
if the model is to describe cochlear function, and therefore must have both
k=0 and a nonzero plate mass, i.e. case A . The actual BM response is
plotted in Fig. 4 for a frequency of 4 kHz . It can be noted that the response
in all three cases is essentially identical until past the peak. Then, case
A dropps off most rapidly. The massless plate, case C , has only a slightly
lower decay rate, while the isotropic plate, case B , has the greatest.

Similar behavior can be found in Fig. 5, the plot of phase vs. distance. Again
cases A and C agree well, while B differs only slightly. If the compliance
is held constant while the BM is considered a clamped, instead of hinged,
plate, the differences for cases A-C are similar in character, though not as
pronounced.

The results obtained here agree with those Steele and Taber (1979) found ana-
lyzing the experimental models of Cannell and Helle. For frequencies in the
physiological range, λL_2 is about 3 at the point of maximum response. For
lower wavenumbers, the dispersion curves for each of cases A , B , and C are
very similar, so it is not surprising that the responses are also similar. At
the point where the signals have decayed to 0 dB , λL_2 is about 25. For larger values the dispersion relations deviate substantially (Fig. 3).

4. ESTIMATE FOR LOCALIZATION

Though the solution of the cochlear model works rather well, the location of
the point of maximum response is difficult to judge without carrying out the
numerical details. In Steele (1974) a cochlear "model 0" is discussed, which
consists of the BM as one of many hinged, tapered strips of an infinite,
massless plate immersed in an infinite fluid. For a given frequency, short
wavelengths occur at points beyond the "transition point" X_{tp} which can be

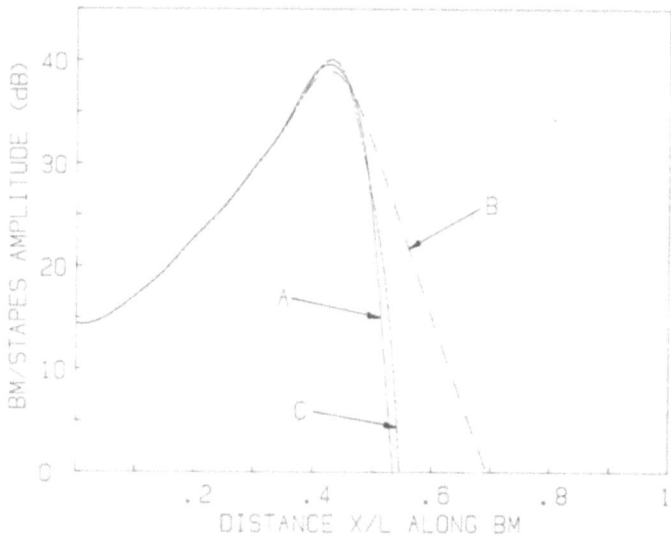

Fig. 4. BM displacement for cases A, B and C . Results are essentially identical up to maximum response. After peak, isotropic plate is most heavily damped, while the orthotropic plate is most lightly. The turning point for this frequency and compliance is X/L = 0.5 .

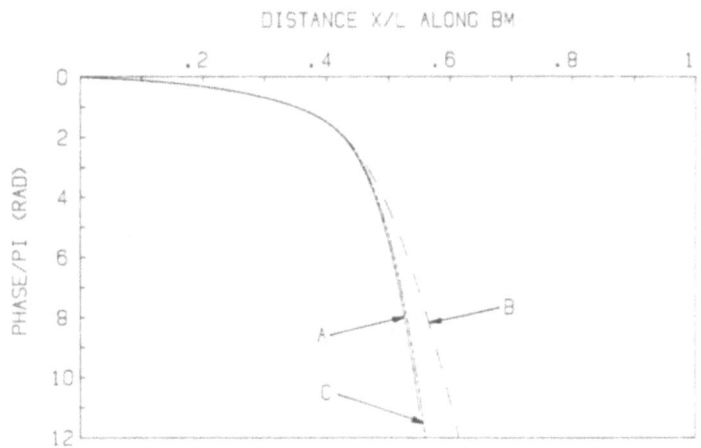

Fig. 5. Phase of traveling wave for cases A , B and C .

defined in terms of the volume compliance there as

$$2\rho\omega^2 = \frac{\pi^5}{120C(x_{tp})} \qquad (2)$$

Comparison with Eq. 1 shows that X occurs at the point at which

$$h_{eq}/b = 1/\pi \qquad (3)$$

At this point, the fluid inertia h_{eq} is near the short wavelength asymptote, $h_{eq} = 1/\lambda$.

Since the damping of the traveling wave becomes severe as the wavelength becomes short, the amplitude is roughly 20 dB down from the peak at X_{tp} . In the guinea pig cochlea for $X_{tp} = 4.1$ mm , the compliance $C=C_B/4=0.15mm^4/N$ gives the frequency f=15kHz, which by coincidence is close to the tuning for the neural fibers at that point. For BM fibers with $E = 210N/mm^2$ and fiber area $A_f=10^{-10}mm^2$, we find

$$C = 8.1b^5/(h^2N_f\gamma) \qquad (4)$$

so that

$$f = 2h(N_f\gamma)^{\frac{1}{2}}/b^{5/2} \qquad (5)$$

where b and h are in mm , f is in kHz , C has the units mm^4/N , and N_f is the number of fibers per mm . The parameter γ is 1,3 or 6 for hinged, constrained, or clamped edges, respectively. The values from various anatomical studies (Tiedemann (1970), Bruns (1976), Cabezudo (1978), Fernandez (1952), and Ehret and Frankenreiter (1977) produce the results in Fig. 6. Details for the filament density N_f are available only for the cat, so this density was assumed for the other cochleas. The width of BMP was assumed to be twice that of BMA.

The correlation of Eq. 5 with the curves for localization obtained by a variety of direct measurements (Bruns (1979), Ehret (1978), Iurato (1962), Wilson and Evans (1977) is excellent for cat, guinea pig, and mouse, while the low-frequency water buffalo might be expected to be similar to the human.

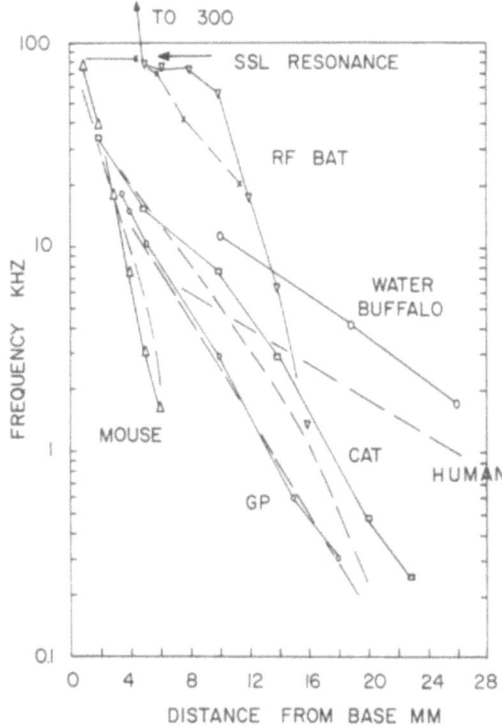

Fig. 6. *Localization of frequency on the basilar membrane. Solid lines show the calculation of the transition point, assuming that the BMP is the only flexible element. Dashed lines indicate results of direct evidence as given in: Rf bat-Bruns (1976), human-Bekesy (1960), cat-Wilson and Evans (1977), guinea pig-Wilson and Johnstone (1975), mouse-Ehret (1978). The cat fiber density was used for all. For the Rf bat, the BMP in the basal 4 mm would flex at 300 kHz; however, the peculiar SSL resonates at 83 kHz .*

Consistent with the anatomy, hinged edges were assumed for the bat, cat and mouse, while constrained edges were assumed for the guinea pig with less pronounced thinning at the edges of BMP . Compared to the measurement C_B in the guinea pig cochlea, Eq. 5 gives about the same in the apical region, and about 25% of C_B in the basal region, which is consistent with the measurements of Dancer and Franke (1980). The curve in Fig. 6 for the water buffalo may be about right, since it is for hinged edges (too flexible) and for the cat N_f (probably too stiff).

The tentative conclusions are:

  (1) BM Compliance can be computed from measurements of the dimensions and microstructure. The fiber density is significant.

  (2) Localization of excitation according to frequency depends primarily on BMP compliance.

ACKNOWLEDGMENT

This work was supported by a grant from the National Institute of Neurological and Communicative Disorder and Stroke.

REFERENCES

Békésy, G. (1960). *Experiments in Hearing* (McGraw-Hill, New York)

Bruns, V. (1976). "Peripheral auditory tuning for fine frequency analysis in the CF-FM bat, Rhinolophus ferremequinum", *J. Comp. Physiol.* 106, 77-97.

Bruns, V. (1979). "Functional anatomy as an approach to frequency analysis in the mammalian cochlea", *Verh. Dtsch. Zool. Ges.* 141-154.

Cabezudo, L. M. (1978). "The ultrastructure of the basilar membrane in the cat", *Acta Otolaryngol* 86, 160-175.

Dancer, A. and Franke, R. (1980). "Intracochlear sound pressure measurements in guinea pigs", *Hearing Research* 2, 190-205.

Ehret, G., and Frankenreiter, M. (1977). "Quantitative analysis of cochlear structures in the house mouse in relation to mechanisms of acoustical information processing", *J. Comp. Physiol.* 122, 65-85.

Ehret, G. (1978). "Stiffness gradient along the basilar membrane as a basis for spatial frequency analysis within the cochlea", *J. Acoust. Soc. Am.* 64, 1723-1726.

Fernandez, C. (1952). "Dimensions of the cochlea (guinea pig)", *J. Acoust. Soc. Am.* 24, 519-523.

Iurato, S. (1962). "Functional implications of the nature and submicroscopic structure of the tectorial and basilar membrane", *J. Acoust. Soc. Am.* 34, 1386-1395.

Lighthill, J. (1981). "Energy flow in the cochlea", *J. Fluid Mechanics* 106, 149-213.

Steele, C. R. (1974). "Cochlear mechanics", *Handbook of Sensory Physiology Vol. V: Auditory System*, eds. W. D. Keidel and W. D. Neff, part 3, 443-478.

Steele, C. R. and Taber, L. A. (1979). "Comparison of WKB and experimental results for three-dimensional cochlear models," *J. Acoust. Soc. Am.* 65, 1007-1018.

Steele, C. R. and Taber, L. A. (1981). "Three-dimensional model calculations for guinea pig cochlea," *J. Acoust. Soc. Am.* 69, 1107-1111.

Tiedemann, H. (1970). "A new approach to theory of hearing", *Acta Otolaryngol. supp.* 277, 1-50,

Wilson, J. P. and Evans, E. F. (1977). "Cochlear frequency map for the cat", *Psychophysics and Physiology of Hearing*, eds. E. F. Evans and J.P. Wilson, Academic Press, London.

ON VIBRATION OF MEMBRANES IN THE MAMMALIAN COCHLEA

V.M. Babič, S.M. Novoselova

V.A. Steklov Mathematical Institute,
The Academy of Sciences of the USSR

ABSTRACT

Three-dimensional hydrodynamical models of the mammalian cochlea are studied. The cochlea is considered as a rectangular long rigid tube, divided into longitudinal canals by either one or two visco-elastic anisotropic plates. One partition is fixed between two rigid slabs and represents the basilar membrane. The second plate can represent either Reissner's membrane (if its edges are fixed) or the tectorial membrane (if one of its edges remains free). The solutions of the models are obtained using Whitham's modification of the WKB method. Calculations on both the two-chambered and the three-chambered model are presented.

1. INTRODUCTION - FORMULATION OF THE PROBLEM

The WKB method is known to be an effective way of theoretical investigation of cochlear mechanics (Steele, 1974a; Steele and Taber, 1979; Babič and Novoselova, 1979; Viergever, 1980). In the present paper we apply the WKB approximation to two- and three-chambered models of the cochlea.

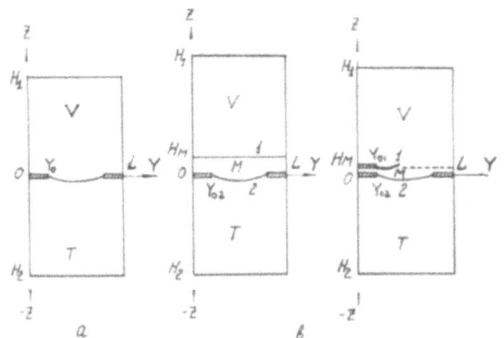

Fig. 1. *Cross-sections of the cochlea models: (a) two-chambered model, (b) three-chambered model with Reissner's membrane (left) and with tectorial membrane (right). V = scala vestibuli, M = scala media, T = scala tympani.*

The cochlea models to be considered have a rectangular cross-section, divided into two or three chambers by flexible partitions (Fig. 1). The motion of the liquids in the canals can be described by the hydrodynamical equations

$$\frac{\partial \vec{v}}{\partial t} = -\frac{1}{\rho} \text{ grad } P + \nu \Delta \vec{v}, \tag{1}$$

$$\text{div}(\rho \vec{v}) = 0, \tag{2}$$

where ρ and ν denote the fluid density and viscosity, and $\vec{v}$ is the velocity of the liquid particles. The boundary conditions on the rigid walls are:

$$\frac{\partial P}{\partial z} = 0 \text{ at } z = H_1, -H_2; \qquad \frac{\partial P}{\partial y} = 0 \text{ at } y = 0, L \tag{3}$$

The normal components of the fluid velocity are continuous across the flexible boundaries. When P^+ and P^- are the pressures just above and below a partition, we have

$$\frac{\partial P^+}{\partial z} = \frac{\partial P^-}{\partial z} \quad \text{at} \quad z = 0, \; H_M. \tag{4}$$

Moreover, we can formulate a relation between the trans-partition pressure $P^- - P^+$ and the partition displacement. We represent the partitions by aniso-tropic plates. The vibration of either partition may then be described by the equation

$$L(W) = \frac{\partial^2}{\partial x^2} (D_x(x,y) \frac{\partial^2 W}{\partial x^2}) + \frac{\partial^2}{\partial x^2} (D_1(x,y) \frac{\partial^2 W}{\partial y^2}) + \frac{\partial^2}{\partial y^2} (D_1(x,y) \frac{\partial^2 W}{\partial x^2}) +$$

$$+ 4 \frac{\partial^2}{\partial x \partial y} (D_{xy} \frac{\partial^2 W}{\partial x \partial y}) + \frac{\partial^2}{\partial y^2} (D_y(x,y) \frac{\partial^2 W}{\partial y^2}) + \mu(x,y) \frac{\partial^2 W}{\partial t^2} = P^- - P^+, \tag{5}$$

where D_x, D_y, D_1, D_{xy} are the components of the anisotropic stiffness, $W(x,y)$ denotes the partition normal displacement, and μ is the surface density of the partition.

The thus formulated boundary value problem (1) - (5) is conveniently solved by means of the WKB approximation.

2. APPROXIMATE SOLUTIONS

The coefficients in Eq. (5) vary relatively slowly with x (that is, they change little within one wavelength). We therefore introduce the 'slow abscissa' $x_1 = \varepsilon x$, where ε is a small parameter. Upon substitution of x_1/ε for x in Eq. (5) we obtain

$$\varepsilon^4 (\frac{\partial^2}{\partial x_1^2} (D_x \frac{\partial^2 W}{\partial x_1^2})) + \varepsilon^2 \{\frac{\partial^2}{\partial x_1^2} (D_1 \frac{\partial^2 W}{\partial y^2}) + \frac{\partial^2}{\partial y^2} (D_1 \frac{\partial^2 W}{\partial x_1^2}) + 4 \frac{\partial^2}{\partial x_1 \partial y} (D_{xy} \frac{\partial^2 W}{\partial x_1 \partial y})\} +$$

$$+ \frac{\partial^2}{\partial y^2} (D_y \frac{\partial^2 W}{\partial y^2}) + \mu \frac{\partial^2 W}{\partial t^2} = P^- - P^+, \qquad D_x = D_x(x_1,y) \ldots \tag{6}$$

The WKB solution for the two-chambered model now has the form (Steele and Taber, 1979; Babič and Novoselova, 1979):

$$P_V = e^{i\theta} \sum_{J=0,1..} P_J \cosh (\zeta_J(H_1-z)) \cos \frac{J\pi y}{L},$$

$$P_T = -e^{i\theta} \sum_{J=0,1..} P_J \cosh (\zeta_J(H_2+z)) \cos \frac{J\pi y}{L}, \tag{7}$$

$$\theta = \omega t - \varepsilon^{-1} \int_0^{x_1} \xi(x_1) dx_1, \qquad -\xi^2 - J^2\pi^2/L^2 + \zeta_J^2 = 0,$$

where P_V and P_T are the pressures in scala vestibuli and scala tympani, respectively, and ξ is the wave number.

For the three-chambered model, a supplementary parameter r is needed to satisfy the boundary condition (4) at the two partitions. The solution then becomes (P_M is the pressure in scala media):

$$P_V = e^{i\theta} \sum_{J=0,1..} P_J \sinh(\zeta_J H_2)\cosh(\zeta_J(r-H_M))\cosh(\zeta_J(H_1-z))\cos\frac{J\pi y}{L} ,$$

$$P_M = e^{i\theta} \sum_{J=0,1..} P_J \sinh(\zeta_J(H_1-H_M))\sinh(\zeta_J h_2)\sinh(\zeta_J(r-z))\cos\frac{J\pi y}{L} , \qquad (8)$$

$$P_T = -e^{i\theta} \sum_{J=0,1..} P_J \sinh(\zeta_J(H_1-H_M))\cosh(\zeta_J r)\cosh(\zeta_J(H_2+z))\cos\frac{J\pi y}{L} .$$

The parameter r has a simple physical meaning in case of non-viscous media. At the surface $z = r$ the pressure P_M is equal to zero and its gradient is directed along the z-axis.

3. DISSIPATION

Following Viergever (1978), we suppose the partitions to be visco-elastic structures. Their anisotropic stiffness then consists of statical and dynamical components:

$$D_y = \hat{D}_y + i\omega R_\nu \frac{h^3}{12} , \qquad D_x = \hat{D}_x + i\omega R_\nu \frac{h^3}{12} , \qquad (9)$$

where $\hat{D}_x$ and $\hat{D}_y$ denote the statical stiffnesses, h is the thickness of the partition and R_ν is some viscous parameter. Perilymph and endolymph have a low viscosity ($\nu = 0.01$ cm^2/sec) so we may take into account only the boundary layer friction. As soon as the acoustical energy condenses near the elastic walls in the acoustical wave-guides, we may neglect the friction at the rigid walls.

Following Inselberg (1978), we introduce supplementary bending moments ΔM_x, ΔM_y, caused by the boundary-layer friction forces:

$$\Delta M_x = \frac{\rho\nu h}{2}\left(\frac{\partial v_x^-}{\partial x} - \frac{\partial v_x^+}{\partial x}\right), \qquad \Delta M_y = \frac{\rho\nu h}{2}\left(\frac{\partial v_y^-}{\partial y} - \frac{\partial v_y^+}{\partial y}\right). \qquad (10)$$

The tangential velocity in an oscillating boundary layer depends on the velocity at infinity as

$$V = V_\infty(1 - e^{-z/\delta}), \qquad \delta^2 = \frac{i\nu}{\omega} . \qquad (11)$$

Introduction of the partial derivatives $\frac{\partial \Delta M_x}{\partial x}$, $\frac{\partial \Delta M_y}{\partial y}$ into Eq. (5) [or Eq. (6)]

yields, with the aid of Eq. (3)

$$L(W) - \frac{\rho \nu h}{2\delta} \left(\frac{\partial V^-}{\partial z} - \frac{\partial V^+}{\partial z} \right) = P^- - P^+. \tag{12}$$

4. SIMPLIFICATION OF THE PARTITION CONDITIONS

We suppose that the partitions have only one transverse mode of displacement. Hence we write

$$W = e^{i\theta} w\eta(y) \tag{13}$$

for each partition velocity $W(x,y)$. For the basilar membrane (BM) and Reissner's membrane (RM), which have fixed edges, an appropriate form of the shape function is

$$\eta(y) = \begin{cases} \sin\left(\frac{\pi y}{b}\right) & \text{for hinged edges,} \\ \sin^2\left(\frac{\pi y}{b}\right) & \text{for clamped egdes,} \end{cases} \tag{14}$$

where b is the membrane width. The tectorial membrane (TM) is represented by a plate with one hinged and one free edge. The corresponding shape function is

$$\eta(y) = \sin(kb)\sinh(ky) + \sinh(kb)\sin(ky), \tag{15}$$

where k is the smallest positive root of

$$\tan(kb) = \tanh(kb). \tag{16}$$

The expressions (8) are rewritten in the form

$$P_V = e^{i\theta} \sum_{J=0,1..} P_{JV} \cosh(\zeta_J(H_1-z)) \cos\frac{J\pi y}{L},$$

$$P_M = e^{i\theta} \sum_{J=0,1..} P_{JM} \sinh(\zeta_J(r-z)) \cos\frac{J\pi y}{L}, \tag{17}$$

$$P_T = e^{i\theta} \sum_{J=0,1..} P_{JT} \cosh(\zeta_J(H_2+z)) \cos\frac{J\pi y}{L}.$$

Adapting the technique of Steele and Taber (1979), we express the coefficients P_{JV}, P_{JM}, P_{JT} in terms of the partition modes:

$$P_{JV} = -\frac{\rho\omega^2 A_{J1} w_1}{\zeta_J \sinh(\zeta_J(H_1-H_M))\delta_J}$$

$$P_{JM} = \begin{cases} -\dfrac{\rho\omega^2 A_{J1} w_1}{\zeta_J \cosh(\zeta_J(r-H_M))\delta_J} \\ \\ -\dfrac{\rho\omega^2 A_{J2} w_2}{\zeta_J \cosh(\zeta_J r)\delta_J} \end{cases} \tag{18}$$

$$P_{JT} = -\frac{\rho\omega^2 A_{J2} w_2}{\zeta_J \sinh(\zeta_J H_2)\delta_J}$$

Here, the subscripts 1 and 2 denote the partition number (see Fig. 1b), and

$$\delta_J = \begin{cases} 1 & \text{if } J = 0, \\ \\ 1/2 & \text{if } J > 0, \end{cases} \qquad A_{Jm} = \frac{1}{L} \int_{y_0}^{y_0+b_m} \eta_m(y-y_0) \cos \frac{J\pi y}{L} \, dy \,, \qquad (19)$$
$$m = 1,2$$

Substitution of the expressions (18) and (17) into Eq. (12) yields as boundary condition at the partitions

$$F = L(e^{i\theta} \eta(y)) -$$

$$e^{i\theta} \sum_{J=0,1..} \frac{\rho\omega^2 A_J^2}{\zeta_J \delta_J} [\tanh(\zeta_J G) + \coth(\zeta_J H)](1 - \frac{\delta h}{2} \zeta_J^2) \cos \frac{J\pi y}{L} = 0,$$
$$(20)$$

where $G = \begin{cases} r & \text{for BM,} \\ H_M - r & \text{for RM or TM,} \end{cases} \qquad H = H_1 - H_M \ (= H_2).$

Equation (20) is simplified by applying the method of softening of the boundary conditions. We equate to zero only the first coefficient of the eigenfunction expansion of the function F. This gives, if stiffness and density do not depend on y,

$$b\{\xi^4 T_1 D_x + 2\xi^2 (\frac{R}{b})^2 T_3 D_3 + (\frac{R}{b})^4 T_2 D_y - \mu\omega^2 T_1\} =$$

$$= \rho\omega^2 L \sum_{J=0,1..} \frac{A_J^2}{\zeta_J \delta_J} [\tanh(\zeta_J G) + \coth(\zeta_J H)] (1 - \frac{\delta h}{2} \zeta_J^2), \qquad (21)$$

where $R = \begin{cases} kb & \text{for TM,} \\ \pi & \text{for BM or RM,} \end{cases} \qquad D_3 = D_1 + 2D_{xy},$

$$T_1 = \frac{1}{b} \int_{y_0}^{y_0+b} \eta^2 dy, \qquad T_2 = \frac{1}{k^4 b} \int_{y_0}^{y_0+b} (\eta'')^2 dy, \qquad T_3 = -\frac{1}{k^2 b} \int_{y_0}^{y_0+b} \eta'' \eta dy.$$

The wavenumber ξ and the parameter r can be calculated from these equations, e.g. by means of Newton iteration.

5. AVERAGED LAGRANGIAN

The evaluation of the amplitude coefficients w_1 and w_2 is based on the averaged Lagrangian method, which was applied by Steele and Taber (1979) for a two-chambered model. For the system with three canals and two partitions, the time-averaged Lagrangian density is

$$L = \sum_{m=1}^{2} \{ \frac{\rho \omega^2 L}{4} \sum_{J=0,1..} \frac{A_{Jm}^2 w_m^2}{\zeta_J^6 \delta_J} [\tanh(\zeta_J G) + \coth(\zeta_J H)] -$$

$$\frac{w_m^2}{4} b_m (\frac{R_m}{b_m})^4 T_{2m} D_{ym} [1-(\frac{\omega}{\omega_m})^2 + \frac{D_{xm}}{D_{ym}}(2 \frac{\xi^2 b_m^2}{R_m^2} \frac{T_{3m}}{T_{2m}} + \frac{\xi^4 b_m^4}{R_m^4} \frac{T_{1m}}{T_{2m}})]\}, \tag{22}$$

where $\omega_m = (\frac{R_m}{b_m})^2 \sqrt{\frac{D_{ym} T_{2m}}{\mu_m T_{1m}}}$, $D_3 = D_x$; $m = 1,2$.

The corresponding Euler equations are

$$\frac{\partial L}{\partial w_1} = 0, \quad \frac{\partial L}{\partial w_2} = 0, \quad \frac{d}{dx} (\frac{\partial L}{\partial \xi}) = 0. \tag{23}$$

The equations $\partial L/\partial w_m = 0$ are equivalent with equation (21) in the absence of dissipation. The third equation of (23) and formulae (18) give the possibility to determine the amplitudes w_m.

6. RESULTS

In our earlier paper (1979) we did not calculate the amplitude coefficients, and the cochlear width was taken equal to the BM width. The present more accurate model has less steep slopes and its maxima are lower (Fig. 2).

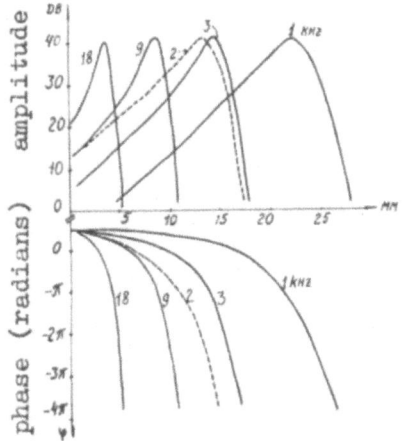

distance from basal end

Fig. 2. BM/stapes transfer ratio of the two-chambered model. Continuous lines denote BM responses to stimuli of 18,9,3,1 kHz. Parameter values:
$\hat{D}_y = D_y = 0.35 \times exp(-x \times 0.127~mm^{-1}) dn \times cm;$
$D_x = 0.001 \times D_y; \nu = 100~cm^2/sec; R_\nu = 0;$
$b = 0.13 \times exp(x \times 0.0438~mm^{-1}) mm;$
$H_1 = H_2 = 1.58 \times exp(-x \times 0.025~mm^{-1}) mm;$
$\mu = 0.0316 \times exp(x \times 0.01~mm^{-1})~g/cm^2.$
The dashed response was calculated with parameter values corresponding to the three-chambered variant, for an input frequency of 2 kHz.

Computations on the three-chambered models showed that the iteration process converges only if the parameter r is close enough to half the distance between the partitions. So, jointly with the boundary conditions at the edge of the two partitions, the stiffness of one plate practically determines the stiffness of the other plate. Therefore we conclude RM to be anisotropic. This anisotropy is consistent with the membrane's small curvature as discussed by Steele (1974b).

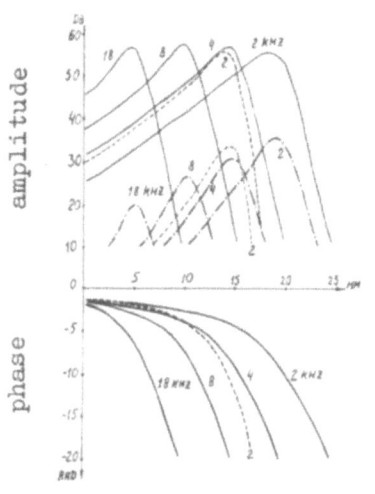

distance from the stapes

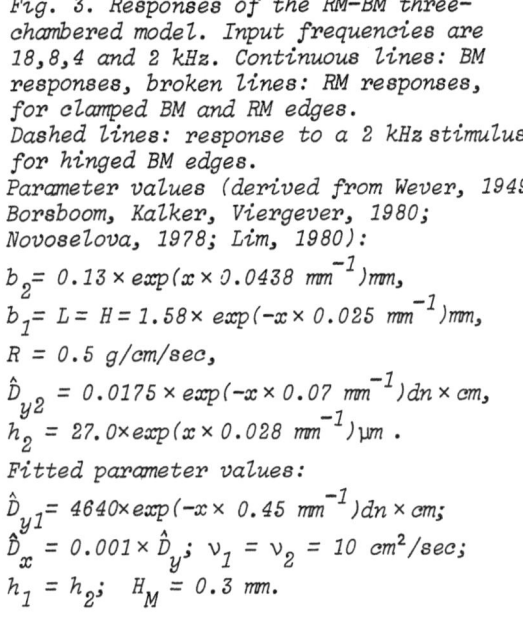

Fig. 3. Responses of the RM-BM three-chambered model. Input frequencies are 18,8,4 and 2 kHz. Continuous lines: BM responses, broken lines: RM responses, for clamped BM and RM edges.
Dashed lines: response to a 2 kHz stimulus for hinged BM edges.
Parameter values (derived from Wever, 1949; Borsboom, Kalker, Viergever, 1980; Novoselova, 1978; Lim, 1980):

$b_2 = 0.13 \times exp(x \times 0.0438\ mm^{-1})mm,$

$b_1 = L = H = 1.58 \times exp(-x \times 0.025\ mm^{-1})mm,$

$R = 0.5\ g/cm/sec,$

$\hat{D}_{y2} = 0.0175 \times exp(-x \times 0.07\ mm^{-1})dn \times cm,$

$h_2 = 27.0 \times exp(x \times 0.028\ mm^{-1})\mu m\ .$

Fitted parameter values:

$\hat{D}_{y1} = 4640 \times exp(-x \times 0.45\ mm^{-1})dn \times cm;$

$\hat{D}_x = 0.001 \times \hat{D}_y;\ \nu_1 = \nu_2 = 10\ cm^2/sec;$

$h_1 = h_2;\ H_M = 0.3\ mm.$

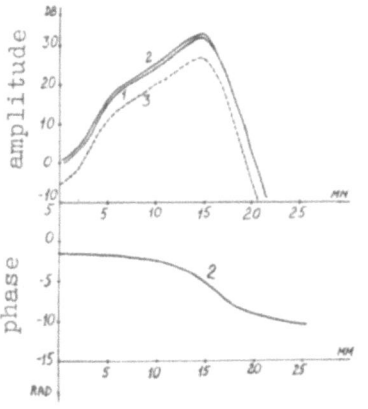

distance from basal end

Fig. 4. Membrane/stapes transfer ratio of the TM-BM three-chambered model.
Curve 1: TM response at $y = y_0 + b_1$.
Curve 2: BM response at the same position.
Curve 3: coincident response of BM and TM at $y = y_0 + b_1/2$. (y_0 is the position of the left edges of BM and TM along the y-axis; b_1, the TM length, equals half the BM width in this model).
Parameter values: input frequency 2 kHz, distance between membranes 5 μm,

$D_{y2} = 5.68 \times 10^4 \times exp(-x \times 0.07\ mm^{-1})dn \times cm.$

Other parameters as in RM-TM model.

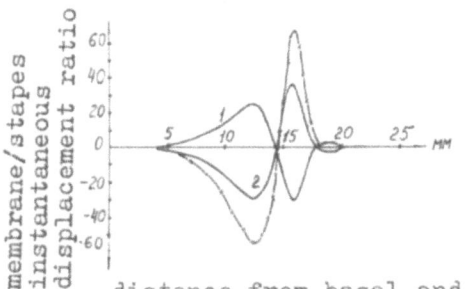

distance from basal end

Fig. 5. Development of the travelling wave in the TM-BM three-chambered model along the line $y = y_0 + b_1$.

Curves 1 and 2 represent the instantaneous displacement of TM and BM. The crossed line shows the altering of distance between the membranes. Parameters as in Fig. 4.

Responses of the model where the upper partition represents RM are shown in
Fig. 3.

If we decrease the thickness of RM and the viscosity near its surface, the
maximum of the RM response moves relatively to the BM maximum. The form of
both tuning curves deteriorates, and double maxima and gaps arise.

Figures 4 and 5 show responses of the model with the upper partition representing the TM. It is remarkable that the TM does not move in the same phase along
its length. This is due to the smallest positive root of Eq. (16) being larger
than π (≈ 3.9). If we put $b_1 = b_2/2$, and the left edges y_{01} and y_{02} of the
partitions are located at y_0, then the line $y = y_0 + b_1/2$ moves in the same
direction as the BM, whereas the line $y = y_0 + b_1$ moves in the opposite direction
(Fig. 5).

REFERENCES

Babic, V.M. and Novoselova, S.M. (1979). On basilar membrane vibrations in the
    mammalian cochlea. In: *Mathematical problems in wave propagation theory*,
    edited by V.M. Babič ("Nauka" Leningrad), vol. 10, pp. 54-62 (in Russian).
Borsboom, M.J.A., Kalker, J.J., and Viergever, M.A. (1980). A geometrically non-
    linear model of the cochlea, *Delft Progress Report* 5, 303-318.
Inselberg, A. (1978). Cochlear dynamics: the evolution of a mathematical model,
    Siam Review 20, 301-351.
Lim, D.J. (1980). Cochlear anatomy related to cochlear micromechanics. A
    review, *J. Acoust. Soc. Am.* 67, 1686-1695.
Novoselova, S.M. (1978). Evaluation of the basilar membrane anisotropic stiff-
    ness, *Biofyzika*, 23, 699-709 (in Russian).
Steele, C.R. (1974a). Behaviour of the basilar membrane with pure-tone exci-
    tation, *J. Acoust. Soc. Am.* 55, 148-162.
Steele, C.R. (1974b). Stiffness of Reissner's membrane, *J. Acoust. Soc. Am.* 56,
    1252-1257.
Steele, C.R. and Taber, L.A. (1979). Comparison of WKB calculations and exper-
    imental results for three-dimensional cochlear models, *J. Acoust. Soc. Am.*
    65, 1007-1018.
Viergever, M.A. (1978). On the physical background of the point-impedance
    characterization of the basilar membrane in cochlear mechanics, *Acustica*,
    39, 292-297.
Viergever, M.A. (1980). *Mechanics of the inner ear*. (Delft University Press).
Wever, E.G. (1949). *Theory of hearing*. (John Wiley and Sons, New York).
Whitham, G.B. (1974). *Linear and nonlinear waves*. (John Wiley and Sons, New
    York).

PSEUDO-RESONANCE IN THE COCHLEA

M. Holmes, J.D. Cole

Rensselaer Polytechnic Institute
Troy, New York   USA

ABSTRACT

A general two-variable approach to the coupled hydroelastic problem of an idealized cochlea is carried out. The basic small parameter is the slenderness of the cochlea. A typical non-linear eigenvalue problem in the transverse cross-section plane results for the phase-function. The slow amplitude and phase variation are obtained analytically. Viscous effects produce traveling waves and a sharp cut-off.

1. INTRODUCTION

It is felt that a simple geometrical model of the cochlea should be adequate for describing its mechanical operation. The essential physical features of incompressible viscous fluid flow and an elastic basilar membrane coupled to it must, however, be included. The general hydroelastic problem is too difficult to be solved analytically or even numerically at present. The basic approach followed here is to exploit the slenderness of the cochlea to construct an asymptotic theory. In this way the problem is reduced to the solution of a mathematical problem in a cross-plane. The general approach is related to that of Steele (1981) and follows in detail that of Chadwick (1980). The theory exhibits traveling waves between the stapes and a (pseudo) resonant point with a rather sharp cut-off.

2. LINEAR HYDROELASTIC PROBLEM

We consider the cochlea to consist of an unrolled tapered tube containing two chambers that are each filled with an incompressible viscous fluid. The planar surface separating the chambers has a rigid section (representing the bony shelf) and a flexible portion (the basilar membrane). For simplicity the cochlear wall is assumed to be symmetric through this plane. Thus, in

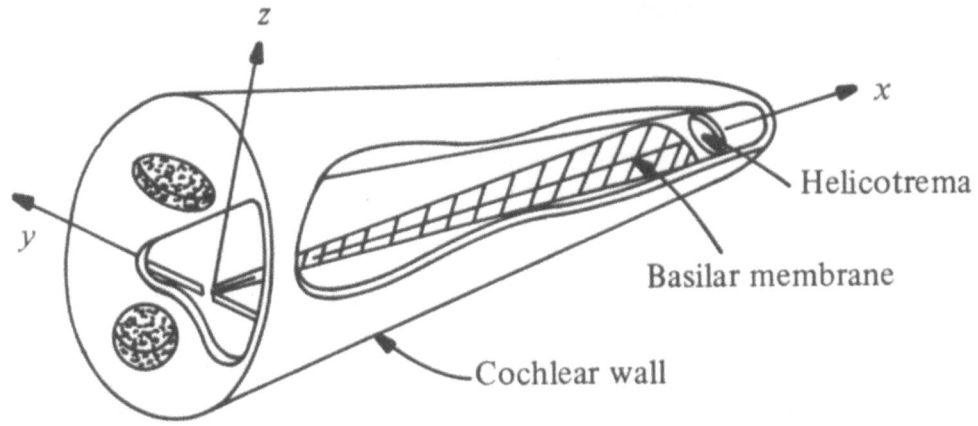

Fig. 1. Geometry and notation for three dimensional hydroelastic model of the cochlea.

studying the response to a pure tone it is only necessary to consider the motion in the upper half of the system.

To describe the dynamical motion in the cochlea we assume the basilar membrane to be an orthotropic elastic plate and the fluid to be Newtonian. The linearized theory of hydro dynamics and elasticity then can be used as the amplitudes are relatively small over a large part of the audible range. Also, since frequencies are greater than 25 Hz, boundary layer theory can be applied to the fluid problem. Therefore, in nondimensional form, the equations of motion for the response to a pure tone (e^{it}) signal are:

    i)   for the fluid pressure $p(x,y,z)$

$$(\epsilon^2\partial_x{}^2 + \partial_y{}^2 + \partial_z{}^2)p = 0 \tag{1a}$$

    ii)   for the basilar membrane deflection $\eta(x,y)$

$$(\partial_y{}^4 + 2D_3\epsilon^2\partial_x{}^2\partial_y{}^2 + D_1\epsilon^4\partial_x{}^4 - \omega^2)\eta = -2\omega^2 p(x,y,0^+). \tag{1b}$$

The parameters are

$$\varepsilon = \frac{B}{L} \qquad \text{and} \qquad \omega = \frac{\Omega}{\Omega_c} \qquad ,$$

where B,L are the width, length of the basilar membrane, respectively, and Ω is the driving frequency. Also, Ω_c is a characteristic resonant frequency of the plate and is given as

$$\Omega_c^2 = \frac{D_2^*}{\mu B^4} \qquad ,$$

where D_2^* is the bending rigidity of the plate in the y-direction and μ is the density of the plate. Finally, D_1 and D_3 are the respective constant bending and twisting rigidities normalized by D_2^*.

The normalized boundary condition for the pressure is

$$(\partial_n - \beta \partial_n^2)p = \begin{cases} -\alpha\eta & \text{on the BM} \\ \\ 0 & \text{on rigid wall} \end{cases} \qquad , \qquad (2)$$

where n is the unit outward normal, $\alpha = \rho B/\mu$,

$$\beta = \sqrt{\frac{\nu}{iB^2\Omega}} \quad ,$$

and ρ, ν are the density and kinematic viscosity of the fluid. As for the plate, it is assumed to be simply supported along its boundary.

It has been pointed out by Dotson (1974) that it may be more appropriate to use a simply supported condition along the spiral ligament, where $y = G_-(x)$, and a clamped condition along the spiral lamina, where $y = G_+(x)$. If these boundary conditions are used there is little qualitative change in the analysis to follow. The simply supported assumption is made primarily to facilitate the discussion. It should also be stressed that due to the variable geometry, the normal in (2) depends on the spatial coordinates and, consequently, on the parameter ε. The constant β in (2) represents the viscous contribution to what is essentially the inviscid problem for the pressure. Since it is complex valued it plays a fundamental role in the attenuation of the wave-like solution that is obtained from (1).

3. SLENDER BODY APPROXIMATION

For the human cochlea, $B \sim 0.5$ mm and $L \sim 3.5$ cm which means that $\varepsilon \sim 10^{-2}$. We can take advantage of this by introducing asymptotic expansions in terms of the small parameter ε. In doing this one finds that the appropriate expansions are of the WKB or two-variable form

$$p \sim \varepsilon \, e^{\frac{i\,\Theta(x)}{\varepsilon}} \left[p_0(x,y,z) + \varepsilon p_1 + \ldots \right] \, , \tag{4a}$$

$$\eta \sim \varepsilon \, e^{\frac{i\,\Theta(x)}{\varepsilon}} \left[\eta_0(x,y) + \varepsilon \eta_1 + \ldots \right] . \tag{4b}$$

Substituting these into (1) and (2) it is found that

$$(\partial_y^2 + \partial_z^2)p_0 = \Theta_x^2 p_0 \, , \tag{5a}$$

$$(\partial_y^4 - 2D_3\Theta_x^2\partial_y^2 + \Theta_x^4 D_1 - \omega^2)\eta_0 = -2\omega^2 p_0 \, , \tag{5b}$$

where

$$(\partial_{n_T} - \beta\partial_{n_T})p_0 = \begin{cases} -\alpha\eta_0 & \text{on BM} \\ 0 & \text{on rigid wall} \, , \end{cases} \tag{6}$$

and along the boundary of the BM, $\eta_0 = \partial_y^2\eta_0 = 0$. In (6), n_T represents the unit outward normal in the transverse cross-section.

Given a reasonably simple geometry one can solve this problem for p_0, η_0, and Θ_x^2. However, it does not determine how p_0 and η_0 vary with the longitudinal variable x, and to determine this it is necessary to consider the $O(\varepsilon^2)$ problem one obtains from substituting (5) into (1). For the case of β small one finds that this leads to the following solvability condition on p_0 and η_0

$$\iint_{\psi} p_o{}^2 dy \ dz + \frac{\alpha}{\omega^2} \int_{G_-}^{G_+} (D_1 \Theta_x{}^2 n_o{}^2 + D_3 n_{o_y}{}^2) dy = \frac{c_o}{\Theta_x} \quad , \tag{7}$$

where c_o is a constant determined from the boundary condition at $x = 0$. The condition at $x = 1$ is not of concern since the wave is damped out exponentially before reaching the distal end for frequencies higher than several hundred Hz.

4. SMALL β APPROXIMATION

With (7) the problem for the first term expansion for small ε is complete. It consists of solving a nonlinear eigenvalue problem (5,6) in each transverse cross-section. After this the slow modulation is determined from (7). Although there are methods to solve this problem, the fact that it is essentially nonlinear complicates the analysis considerably. It can be simplified somewhat, as is done below, by reintroducing the boundary layer approximation used to obtain (2).

Recalling that for the audible spectrum $\beta < < 1$ then

$$\Theta_x(x) \sim k_o(x) + \beta k_1(x) \ . \tag{8}$$

From (5,6), to the first order in β, we obtain the inviscid cross-plane problem

$$(\partial_y{}^2 + \partial_z{}^2) p_o = k_o{}^2 p_o \tag{9a}$$

$$(\partial_y{}^4 - 2D_3 k_o{}^2 \partial_y{}^2 + k_o{}^4 D_1 - \omega^2) n_o = -2\omega^2 p_o(x,y,0) \quad , \tag{9b}$$

and

$$\partial_{n_T} p_o = \begin{cases} -\alpha n_o & \text{on BM} \\ \\ 0 & \text{on rigid wall} \end{cases} \tag{10}$$

The viscous correction to k_o in (8) is found from the $O(\beta)$ problem that comes from (5,6) and is given as

$$k_1 = \frac{k_0 \int_{\partial\Psi} p_0^2}{2\iint_\Psi p_0^2 \, dy \, dz + 2\alpha/\omega^2 \int_{G_-}^{G_+} (D_1 k_0^2 \eta_0^2 + D_3 \eta_{o_y}^2) \, dy} \tag{11}$$

Thus, one only needs to solve the somewhat simpler real eigenvalue problem (9,10), then determine the viscous correction to the phase function from (11).

5. SOLUTION OF NONLINEAR EIGENVALUE PROBLEM

There are a number of ways to solve (9,10) for the first term expansions of the fluid pressure and plate displacement. For example, one could use numerical methods, Green's functions, or modal expansions. Each has its restrictions as well as its advantages. To illustrate one method consider the special idealized case of a rectangular cross-section as shown in Fig. 2.

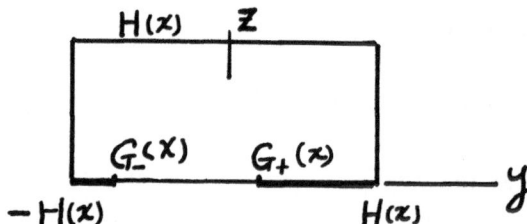

Fig. 2. Transverse cross-section used for modal expansion.

Separating variables in (9a) and substituting the result into (9b) one finds that

$$(\partial_y^4 - 2D_3 k_0^2 \partial_y^2 + D_1 k_0^4 - \omega^2)\eta_0 = \sum_{m=0}^{\infty} a_m \cos\gamma_m y \int_{G_-}^{G_+} \eta_0 \cos\gamma_m s \, ds \tag{12}$$

where

$$a_m = \frac{2\alpha\omega^2 c_m}{\lambda_m} \coth\lambda_m H \quad, \qquad c_m = \begin{cases} 1/2H \;, & m = 0 \\ 1/H \;, & m \neq 0 \end{cases}$$

$$\lambda_m^2 = \gamma_m^2 + k_0^2 \qquad , \qquad \gamma_m = \frac{m\pi}{H} \; .$$

The nonlinear dependence on k_o^2 is clearly seen in this result and to handle it we also expand η_0 in modes as follows

$$\eta_0 = \sum_{\ell=1}^{\infty} b_\ell \cos \lambda_\ell (y - G_-) \quad , \quad \lambda_\ell = (2n - 1)\frac{\pi}{2} (G_+ - G_-) .$$

From this and (12) one obtains the following algebraic problem for k_o^2 and b_ℓ

$$(\lambda_\ell^4 + 2D_3 \lambda_\ell^2 k_o^2 + D_1 k_o^4 - \omega^2) b_\ell = \sum_{m=0}^{\infty} \sum_{n=1}^{\infty} a_m b_n K_{mn} K_{m\ell} \tag{13}$$

where

$$K_{mn} = \frac{2}{G_+ - G_-} \int_{G_-}^{G_+} \cos \gamma_m y \cdot \cos \lambda_n (y - G_-) dy$$

Although there are an infinite number of equations for the b_ℓ's in (13), the first few terms should serve as a reasonable approximation to the solution. From this the complete solution is found by determining k_1 in (11) and by also satisfying (7).

6. NUMERICAL EXAMPLE

As an example of the above analysis we now consider the specific case $B = 0.05$ cm, $L = 3.5$ cm, $\nu = 0.008$ cm/sec, $\rho = 1.0$ gm/cm^2, and

$$D_2^* = \frac{Eh^3}{12(1 - \sigma^2)} ,$$

where

$$E = 4 \times 10^6 \text{ dyn/cm}^2 \quad , \quad h = 1.05 \times 10^{-3} \text{ cm} \quad , \quad \text{and } \sigma = 0.5 .$$

Also it is assumed that the plate is highly orthotropic so

$$D_1 = D_3 = 0 \quad , \quad H = 1.5 \quad , \quad \text{and } G_+ = -G_- = G(x) \text{ where}$$

$$G(x) = \frac{1}{12} (5x + 1) \quad , \quad 0 < x < 1 .$$

The result for the case of $\ell = 1$ and $m = 5$ in (13) is shown in Fig. 3, where we have taken $x = 0.5$. It is clear from this preliminary calculation

52

that the tuning is relatively sharp.

In general, if k_0 is real then waves exist; if k_0 is imaginary the solution damps (or grows) exponentially. The effect of viscosity in k_1 causes the standing waves of the zero viscosity solutions to become traveling waves, and also damps highly the short waves near the pseudo-resonant point.

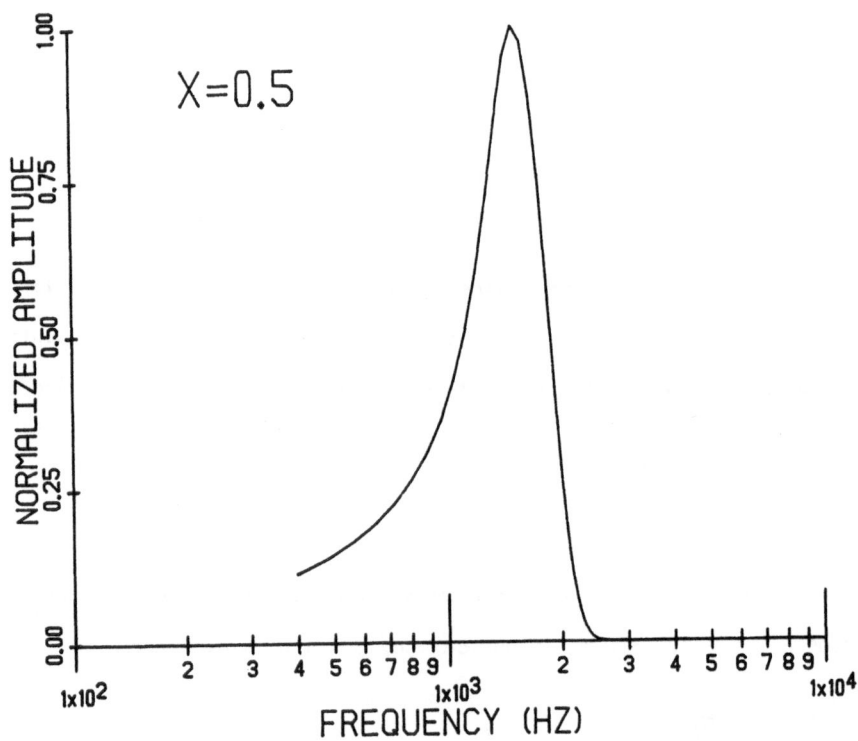

Fig. 3. Tuning curve obtained at the longitudinal location $x = 0.5$.

REFERENCES

Chadwick, R.S. (1980). Studies in Cochlear Mechanics. In: *Mathematical Modeling of the Hearing Process*, Lecture Notes in Biomathematics No. 43, Springer-Verlag, Berlin.
Dotson, R. (1974). *Transients in a Cochlear Model*, PhD. thesis, Stanford University.
Steele, C.R. (1981). *Lecture Notes in Cochlear Mechanics*, SIAM Regional Conference R.P.I., 1980.

SIMULTANEOUS AMPLITUDE AND PHASE MATCH OF COCHLEAR MODEL CALCULATIONS AND BASILAR MEMBRANE VIBRATION DATA

Max A. Viergever, Rob J. Diependaal

Department of Mathematics & Informatics
Delft University of Technology, The Netherlands

ABSTRACT

One of the major problems in cochlear mechanics has been the incapability of cochlea models to provide a simultaneous match to amplitude and phase data of basilar membrane (BM) vibration. Negative results were reported by Allen and Sondhi (1979), Viergever (1980), and Neely (1981), all using a two-dimensional (2D) cochlea model. Steele and Taber (1981) on the other hand found fair agreement between their 3D model calculations and Wilson and Johnstone's (1975) BM data, which is puzzling since the model parameters were in the range where 2D and 3D responses do not differ significantly.
The apparent contradiction is mainly due to an incorrect interpretation of 2D model results. The 'BM velocity' in the customary 2D model is the average of the partition velocity over the channel width rather than the average over the membrane width or the velocity of the BM centreline. The BM/stapes amplitude ratio has, consequently, been underestimated by 10-30 dB in most 2D model calculations. By using the correct definition we have reached good agreement between measurement results (Rhode 1971; Johnstone and Yates 1974) and 2D model results as regards both amplitude and phase.

1. INTRODUCTION - STATEMENT OF THE PROBLEM

Although the focus of mathematical modelling of cochlear mechanics has recently shifted towards nonlinear and active processes, the behaviour of linear, passive models is not yet fully understood. For the past few years the two main problems have been (i) the phenomenon of wave reflection, or rather the almost complete absence of it, and (ii) the insufficient quantitative agreement of model calculations and experimental data of basilar membrane (BM) vibration. These problems now approach their solution. The absence of wave reflection in passive cochlea models is amply discussed in two forthcoming articles (De Boer, 1983; De Boer and Viergever, 1983), while the present paper deals with the comparison of model results and measurement results.

The validation of macromechanical (linear and passive) cochlea models against experimental observations of BM motion has shown that the models are quite acceptable in a qualitative sense, but are susceptible of improvement in a quantitative sense. In particular, it has appeared to be possible to match either the amplitude or the phase of the BM/stapes transfer ratio, but not both simultaneously, i.e. with one set of parameters. This conclusion has been reached by Allen and Sondhi (1979) and Neely (1981) on the basis of Rhode's (1971) squirrel monkey data, and by Viergever (1980) on the basis of guinea pig data recorded by B.M. Johnstone and Yates (1974) and by Wilson and J.R. Johnstone (1972, 1975). The cochlea model used by all authors was two-dimensional (2D).

In contrast with these findings, Steele and Taber (1981) did report good agreement between their model results and Wilson and Johnstone's (1975) measurement results. They used a 3D model in their calculations. The obvious explanation of the discrepancy with the 2D model studies would be that a 3D model is significantly more accurate in simulating the BM response than a 2D model. This is not true, however. The fluid pressure in a 3D model depends strongly on the transverse coordinate, so fluid motion is fully three-dimensional. Nevertheless, 2D and 3D models agree to a large extent as regards BM motion (Steele and Taber, 1979b). Hence, the discrepancy must have another origin.

We discovered the reason for the failure of the 2D calculations when we compared 1D, 2D and 3D responses of cochlear models (Viergever and Diependaal, 1983). Each 2D model is, in a more or less explicit manner, derived from a 3D geometry by omitting fluid pressure variations in the direction lateral to the BM. This implies that the pressure in a 2D model is an average of the actual (3D) pressure over the channel width. Consequently, by applying the relation P = ZV, where P is the 2D transmembrane pressure and Z is the BM impedance, a velocity V is obtained which is not a BM velocity, but the average over the channel width of the velocity of the cochlear partition. Since the partition velocity is identical to zero except for the part covered by the BM, this procedure underestimates the model response by a factor of b/β, the ratio of channel width to BM width, if the average of the BM velocity over its width is the desired output, or by a factor of $\pi b/2\beta$ if one is interested in the velocity of the BM centreline. Allen and Sondhi (1979), Viergever (1980), and Neely (1981) did not take this effect into account, which is why their 2D calculations could not be brought in agreement with experimental observations.

The aim of the present study is to demonstrate that a good match to the BM vibration data used in the mentioned 2D model studies can be accomplished as regards both amplitude and phase. Our starting point is a 3D box model of the cochlea, for which an approximate 2D response is formulated based on the Liouville-Green (LG) method. We have opted for a 2D solution (instead of a 3D one) because the LG approximation is less trustworthy in the 3D case (De Boer and Viergever, 1982). The model results are matched to data of Johnstone and Yates (1974) and Rhode (1971) by means of a curve fitting procedure with a limited number of free parameters. The results justify the conclusion that the type of cochlear model considered (i.e., linear and passive) adequately describes the selected experimental findings.

2. LIST OF SYMBOLS

x,y,z	coordinates of the cochlea model, see Fig. 1
b,h,l	width, height, length of a chamber of the model
$\beta(x)$	BM width
ω	radian frequency of the stapes motion
ρ	fluid density
σ	density of the BM material
$H(x)$	thickness of BM plus attached cells
$D(x)$	flexural rigidity of the visco-elastic beams
$D_1(x)$, $D_2(x)$	components of $D(x)$
$Z(x)$	specific acoustic impedance of the BM
$M(x)$, $R(x)$, $S(x)$	mass, resistance, stiffness of the BM per unit area
$V(x)$	velocity of the BM centreline normalized to the stapes velocity
$k(x)$	wave number of the BM velocity wave
$Q(k)$	geometry function
j	imaginary unity
x_{obs}	point of observation

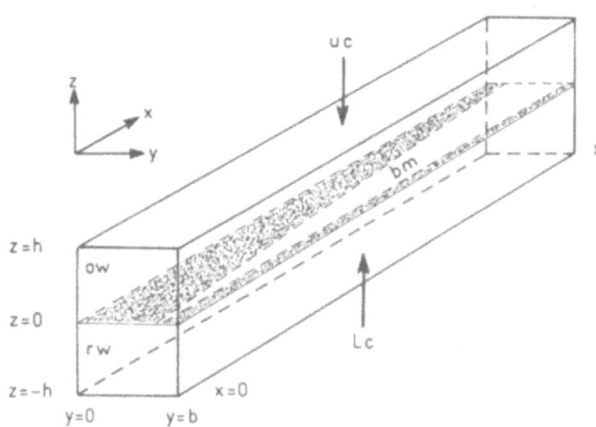

Fig. 1. Geometry of the cochlea model.

bm: *basilar membrane*
uc: *upper chamber*
lc: *lower chamber*
ow: *oval window*
rw: *round window*

3. MODEL AND SOLUTION METHOD

The cochlea is modelled as a straight, two-chambered box, see Fig. 1. The walls of the box are rigid, with the exception of those at the basal end (x=0), which represent the oval window and the round window. The fluid in the two chambers is assumed to be incompressible and inviscid, and to behave linearly. The partition, located at z=0, has a rigid part (the shaded area in Fig. 1) and a flexible part, the basilar membrane. The BM is represented by a series of parallel linearly visco-elastic simply supported beams in the transverse direction (the y-direction). These simplifications, as well as minor ones which were not mentioned here, have been justified in Viergever (1980, chapters 2 and 3). The equations describing the movements of the fluid and the BM to stimulation by the stapes (via the oval window) can be found in the same reference.

The model equations can be used for 1D, 2D and 3D calculations. The 3D mode is
obtained by allowing the fluid pressure to vary in all three spatial dimensions.
In the 2D mode the pressure varies in the x- and z-directions, but not in the
y-direction, whereas in the 1D mode only variations in the x-direction are
taken into consideration. The BM velocity is assumed to have a half sine-shaped
distribution over the membrane width in all modes; there are no additional assumptions concerning this quantity in the 1D and 2D cases.

The appropriate dimensionality of the calculations depends on the intent of the
study. It has been shown repeatedly that the 1D approximation is adequate only
for qualitative purposes, hence it is not suited for the present work. The
choice between 2D and 3D is more difficult. The 3D mode is, of course, slightly
more accurate on account of the extra space dimension, but there is a conflict
with the requirement that we need to have a solution technique that is both reliable and fast. Computational speed is important for two reasons. First, the
principal model parameters are known only by order-of-magnitude estimates, which
necessitates extensive parameter variation in fitting the measurement data.
Second, the output of the model is BM velocity as a function of the longitudinal
coordinate x, for a fixed input frequency, whereas the data are recorded in the
form of frequency response curves, that is a response at a fixed point on the
BM as a function of stimulus frequency. Consequently, a comparison between the
two requires solution of the model equations for a large number of frequencies.

The model equations are so complicated, particularly owing to the intricate
structure of the cochlear partition, that they do not admit an analytic solution.
Approximating the solution by a straightforward numerical technique has neither
been feasible in the 3D mode because of computer storage problems. For the 2D
case numerical solutions have been obtained (Allen, 1977; Allen and Sondhi,
1979; Viergever, 1980; Neely, 1981), but the long computation times preclude
numerical experimentation with the parameters. We must, consequently, settle for
an asymptotic approach. The most suitable asymptotic method for solving cochlear
mechanics problems is the LG approximation, which is based on the assumption
that the BM wave travels in a medium of which the propagation properties do not
vary much within one wavelength. The method has proved to be fairly reliable
(Steele and Taber, 1979a; Viergever, 1980), although it has several pitfalls
that are difficult, if at all, to avoid (De Boer and Viergever, 1982). An additional feature of the method is that its performance is best in one dimension
and worst in three. Especially the decay of the amplitude envelope beyond the
peak is too steep in the 3D response. It is, therefore, questionable whether
2D or 3D calculations are to be preferred: 3D is more accurate than 2D, but

the LG approximation is better for 2D than for 3D. We have, somewhat arbitrarily
chosen for the 2D mode.

The LG solution to the cochlear model of Fig. 1. is (Steele and Taber, 1979b;
Viergever and Diependaal, 1983)

$$V(x) = - \frac{j\pi bhk(0)}{2\beta(0)} \left[\frac{\beta dQ/dk|_{x=0}}{\beta dQ/dk} \right]^{\frac{1}{2}} \exp\{-j \int_0^x k(\xi)d\xi\}, \qquad (1)$$

with Q(k) satisfying

$$Q(k) = - \frac{Z(x)}{2j\omega\rho}. \qquad (2)$$

For an explanation of the symbols, see Section 2.
The impedance Z(x) of the BM is related to the parameters of the visco-elastic
beam system by

$$Z(x) = j\omega\sigma H(x) + \frac{\pi^4 D(x)}{j\omega\beta^4(x)}. \qquad (3)$$

We suppose that the BM consists of Kelvin material, the simplest visco-elastic
material for a solid (Flügge, 1975, p.9), which implies that D(x) has the form
$D = D_1 + j\omega D_2$, with D_1 and D_2 real-valued quantities. Then Eq. (3) can be
written in the form

$$Z(x) = j\omega M(x) + R(x) + \frac{S(x)}{j\omega}. \qquad (4)$$

The function Q(k) depends on the dimensionality of the fluid flow in the model
and on geometrical parameters as chamber height and ratio BM width/chamber
width. For the 2D mode of the model of Fig. 1, the value of Q(k) is

$$Q(k) = \frac{8\beta}{\pi^2 bk \tanh(kh)}. \qquad (5)$$

The 2D LG solution for the BM/stapes transfer ratio thus becomes

$$V(x) = - \frac{j\pi bhk(x)}{2\beta(x)} \left[\frac{hk(0)\,\text{csch}^2\{hk(0)\} + \coth\{hk(0)\}}{hk(x)\,\text{csch}^2\{hk(x)\} + \coth\{hk(x)\}} \right]^{\frac{1}{2}} \times$$
$$\times \exp\{-j \int_0^x k(\xi)d\xi\}, \qquad (6)$$

with k(x) to be solved from

$$k \tanh(kh) = - \frac{16j\omega\rho\beta}{\pi^2 bZ}. \qquad (7)$$

4. RESULTS

Figure 2 shows a comparison of our calculations with guinea pig measurements of Johnstone and Yates, and Fig. 3 with data recorded by Rhode in the squirrel monkey. A few remarks need to be made so as to explain how the figures were produced.

- Both experimental results were obtained with the Mössbauer technique. The Mössbauer source covers a large part of the BM width, so the measured response will be an average in the lateral direction rather than the velocity of the centreline. We estimated that the data represent an average over 2/3 of the width of the BM. In our model the BM velocity has a half sine shaped distribution over the width. Hence we multiplied $V(x)$ by a factor $3\sqrt{3}/2\pi$, which amounts to a reduction of 1.65 dB.
- The data in Fig. 3, Rhode's 69-473 squirrel monkey results, were taken from the paper of Zweig, Lipes and Pierce (1976), since Rhode (1971) only published the amplitude of the response. Furthermore, Rhode measured BM/malleus transfer functions, while the cochlea model yields a BM/stapes ratio. We therefore adapted Rhode's data in conformity with his (1978, Fig. 5) stapes/malleus transfer ratios. The resulting reduction of the peak of the amplitude curve is consistent with Rhode's own findings.
- Johnstone and Yates did not supply the point of observation on the BM in their study. We estimated it to be at 3 mm from the basal end using Wilson and Johnstone's (1975, Fig. 22) plot of cutoff frequency against position along the membrane. Rhode neither supplied the observation point. A reliable cochlear map is not available for the squirrel monkey, so we rather arbitrarily set x_{obs} = 15 mm in Fig. 3. We have checked that different choices of x_{obs} did not significantly affect the quality of the fits. The values of the parameters M_0, R_0 and S_0 (see below) appeared to be quite sensitive to changes in x_{obs}, however.
- Several of the model parameters were kept fixed in the calculations, viz. b, h, ρ and $\beta(x)$. The values for guinea pig were derived from Fernández' (1952) data, those for squirrel monkey were estimated since measurements of geometrical parameters in this species are not known to us. The values used can be found in the legend to the figures. Notice that the length of the cochlea is irrelevant. It suffices to consider the interval $0 \leq x \leq x_{obs}$, because the response in x_{obs} is uninfluenced by the part of the model with $x > x_{obs}$, cfr. Eq. (6). This is a consequence of the unidirectionality of the LG approach.
- The remaining parameters M, R and S were written in the form $M_0\exp(M_1 x)$, $R_0\exp(R_1 x)$, $S_0\exp(S_1 x)$. The exponents M_1, R_1 and S_1 can be estimated since

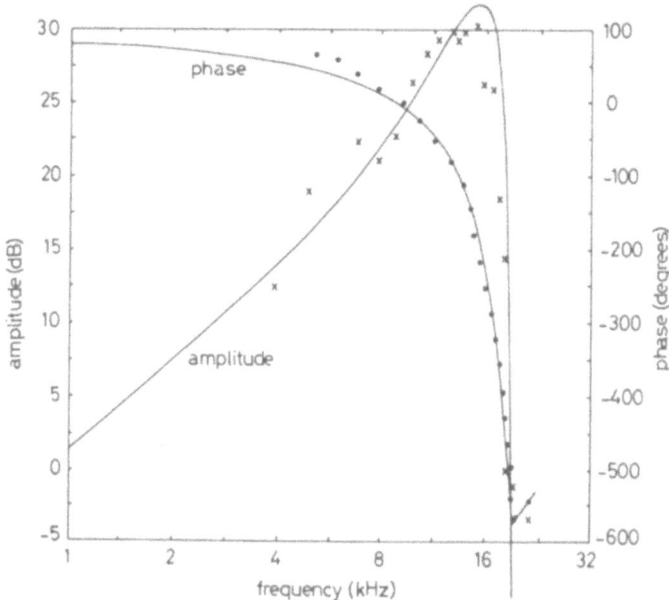

Fig. 2. Comparison of model results with data (crosses: amplitude, dots: phase)
observed in the living guinea pig by Johnstone and Yates (1974, Fig. 3). The
quantity displayed is the BM/stapes transfer ratio. Parameter values: x_{obs} =
3 mm, b = 0.5 mm, h = 1.4 mm, ρ = 1.0 mg/mm^3, β = 0.08 exp (0.04x) mm,
M = 0.098 mg/mm^2, R = 3.8 exp(-0.275x) μNs/mm^3, S = 7.7 exp(-0.55x) N/mm^3.

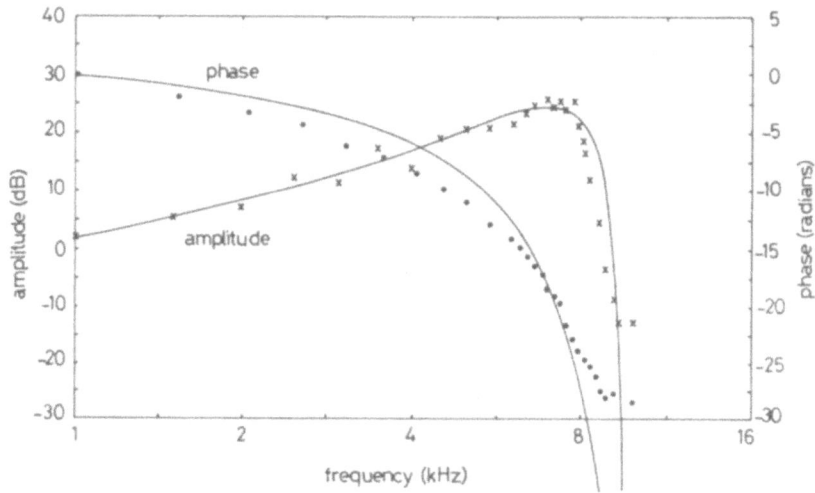

Fig. 3. Comparison of model results with Rhode's in vivo 69-473 squirrel monkey
data (crosses: amplitude, dots: phase). The data were transformed to BM/stapes
ratios using Rhode's (1978, Fig. 5) stapes/malleus transfer function. Parameter
values: x_{obs} = 15 mm, b = 0.5 mm, h = 1.0 mm, ρ = 1.0 mg/mm^3, β = 0.08 exp
(0.05x) mm, M = 0.09 mg/mm^2, R = 1.2 exp(-0.10x) μNs/mm^3, S = 7.1 exp(-0.20x)
N/mm^3.

they relate to geometrical parameters (see Viergever, 1980, Section 5.1). The coefficients M_0, R_0 and S_0 are known only roughly; they were determined by a curve fitting procedure. The resulting values are shown in the legends to Figs. 2 and 3.

5. DISCUSSION OF THE RESULTS

The agreement between our 2D model calculations and the measurement results of Johnstone and Yates, as shown in Fig. 2, is excellent. The only discrepancy is that the amplitude plateau of the model response is lower than that of the measurements (it occurs at -19 dB), but this is a typical shortcoming of the LG approach (Viergever, 1980, Fig. 5.2.4). The match to Rhode's data in Fig. 3 is less good, but still quite acceptable considering that BM motion was nonlinear in Rhode's observations. Here both the amplitude plateau and the phase plateau of the model are much lower than those of the measurements, again as a result of the LG approximation.

Steele and Taber (1981) were the first to bring cochlear model calculations in fair quantitative agreement with experimental results of BM motion. They did so by comparing a 3D LG solution for a model similar to that of Fig. 1 with guinea pig data of Wilson and Johnstone (1975). The extent to which the calculated and measured responses agree is remarkable inasmuch as Steele and Taber did not make use of a curve fitting procedure.

The results of the present study, together with those of Steele and Taber, show that the simple box model of Fig. 1 with linear and passive BM characteristics fully serves its purpose. It adequately describes the measured linear response of the BM to stapes movements. This is a quite satisfactory conclusion of our studies in cochlear macromechanics. It also justifies some optimism as regards the challenge offered to cochlear modelling by the recent observations of Khanna and Leonard (1982) and Sellick, Patuzzi, and Johnstone (1982), which demonstrate that the nonlinear and locally active processes that take place in the intact cochlea clearly manifest themselves at the level of BM vibration.

REFERENCES

Allen, J.B. (1977). Two-dimensional cochlear fluid model: New results, *J. Acoust Soc. Am.* 61, 110-119.
Allen, J.B. and Sondhi, M.M. (1979). Cochlear macromechanics: Time domain solutions, *J. Acoust. Soc. Am.* 66, 123-132.

Boer, E. de (1983). Auditory Physics. Physical principles in hearing theory Il.
    To appear in *Phys. Rep.*
Boer, E. de and Viergever, M.A. (1982).Validity of the Liouville-Green (or WKB)
    method for cochlear mechanics, *Hearing Res.* 8, 131-155.
Boer, E. de and Viergever, M.A. (1983). Wave propagation and dispersion in the
    cochlea. Submitted to *Hearing Res.*
Fernández, C. (1952). Dimensions of the cochlea (guinea pig), *J. Acoust. Soc.
    Am.* 24, 519-523.
Flügge, W. (1975). *Viscoelasticity,* Springer, Berlin.
Johnstone, B.M. and Yates, G.K. (1974). Basilar membrane tuning curves in the
    guinea pig, *J. Acoust. Soc. Am.,* 55, 584-587.

Khanna, S.M. and Leonard, D.G.B. (1982). Basilar membrane tuning in the cat
    cochlea, *Science,*215, 305-306.
Neely, S.T. (1981). Finite difference solution of a two-dimensional mathematical
    model of the cochlea, *J. Acoust. Soc. Am.* 69, 1386-1393.
Rhode, W.S. (1971). Observations of the vibration of the basilar membrane in
    squirrel monkeys using the Mössbauer technique, *J. Acoust. Soc. Am.* 49,
    1218-1231.
Rhode, W.S. (1978). Some observations on cochlear mechanics, *J. Acoust. Soc.
    Am.* 64, 158-176.
Sellick, P.M., Patuzzi, R., and Johnstone, B.M. (1982). Measurement of basilar
    membrane motion in the guinea pig using the Mössbauer technique, *J. Acoust.
    Soc. Am.* 72, 131-141.
Steele, C.R. and Taber, L.A. (1979a). Comparison of WKB and finite difference
    calculations for a two-dimensional cochlear model, *J. Acoust. Soc. Am.* 65,
    1001-1006.
Steele, C.R. and Taber, L.A. (1979b). Comparison of WKB calculations and exper-
    imental results for three-dimensional cochlear models, *J. Acoust. Soc. Am.*
    65, 1007-1018.
Steele, C.R. and Taber, L.A. (1981). Three-dimensional model calculations for
    guinea pig cochlea, *J. Acoust. Soc. Am.* 69, 1107-1111.
Viergever, M.A. (1980). *Mechanics of the inner ear - a mathematical approach.*
    Delft University Press.
Viergever, M.A. and Diependaal, R.J. (1983). Liouville-Green solutions of one-,
    two- and three-dimensional cochlea models. In preparation.
Wilson, J.P. and Johnstone, J.R. (1972). Capacitive probe measures of basilar
    membrane vibration. In: *Hearing theory 1972,* IPO Eindhoven, 172-181.
Wilson, J.P. and Johnstone, J.R. (1975). Basilar membrane and middle-ear vi-
    bration in guinea pig measured by capacitive probe, *J. Acoust. Soc. Am.*
    57, 705-723.
Zweig, G., Lipes, R., and Pierce, J.R. (1976). The cochlear compromise, *J.
    Acoust. Soc. Am.* 59, 975-982.

ADVANTAGES FROM DESCRIBING COCHLEAR MECHANICS IN TERMS OF ENERGY FLOW

J. Lighthill

University College London
United Kingdom

ABSTRACT

The interpretation of observations of basilar-membrane response to pure tones,
and also the analysis of mathematical models of the mechanical response of the
cochlea, are both valuably facilitated by a description in terms of energy
flow. An attempt is made to describe this approach clearly, to demonstrate its
advantages, and to relate it to other approaches.

1. INTRODUCTION

Insight into the extent of the contribution from cochlear mechanics to auditory
sensitivity in frequency discrimination has to be derived by combining information from two sources. First, an inevitably limited range of in-vivo observations of basilar-membrane response to pure tones at various locations on the
membrane is (in certain species) available. Secondly, many mathematical models
of the cochlea's mechanical response to pure tones are available for various
assumed mechanical properties and at various levels of modelling complication.

Under these circumstances, little is gained by simply looking for a mathematical model whose results are in general agreement with observation in those
cases where reliable in-vivo observations are available. It is necessary to
look for models that successfully bridge the gap between (a) the regrettably
far from accurate knowledge of in-vivo mechanical properties and (b) the
insufficiently extensive observations of cochlear response. Only models
consistent both with (a) and with (b) can be provisionally used to extend
knowledge beyond its current limitations.

A model must be regarded as failing to bridge that gap if it demands unrealistic values of geometrical or mechanical parameters important for the cochlea's
mechanical response. There is, in fact, a special reason why a wide range of
models making seriously oversimplified assumptions on matters that significantly influence the mechanics may give misleadingly 'good' predictions. This
reason is that the 'critical-layer absorption' property common (see Lighthill
1981, and section 3 below) both to several relatively realistic models and to
many models of varying degrees of unrealism (serious as in two-dimensional
models or gross as in one-dimensional) does already suffice to ensure

prediction of the very sharp high-frequency cut-off characteristic of in-vivo
basilar-membrane response measurements.

On the other hand, a model utilising a realistic degree of relevant mechanical
and geometrical complexity may require such an immense processing exercise
from the computer that only its overall conclusions can be compared with
observations; while detailed insight into how the model's operation at each
location is related to assumptions about local properties may be lacking.
Fortunately, however, it is well established that high-frequency asymptotics,
under a wide variety of different names (see for example Steele 1974, De Boer
1979, Viergever 1980, Holmes 1982) succeeds in avoiding this difficulty while
(for all the frequencies of principal interest for cochlear mechanics) giving
results that agree very well with results of accurate computations (Steele and
Taber 1979).

Although most of those writers who have usefully made complicated cochlear-
mechanics models tractable through high-frequency asymptotics regarded it as,
essentially, a mathematical device, the purposes of a model can be still
better served if we utilise the well established one-to-one relationship between such asymptotics and the simple physical principles of energy flow. It
is known (see for example Whitham 1974 or Lighthill 1978) that high-frequency
asymptotics, applied to analyse vibrating systems, gives results identical
with those obtained by making certain assumptions on how vibrational energy is
changing as a result of energy flow and energy attenuation; the velocity of
energy flow (or 'group velocity') being given as the gradient of a plot of
frequency against wavenumber. This way of expressing the results from a model
is summarised, for the cochlear-mechanics application, in the next section.
While mathematically equivalent to high-frequency asymptotics, it has the
advantage of a simple physical interpretation, yielding insight into how the
model's operation at each location is related to assumptions about local
properties.

2. ENERGY-FLOW DESCRIPTION OF COCHLEAR RESPONSE TO PURE TONES

The fundamental assumption, additional to that of high frequency, which under-
lies both the mathematical asymptotics and their equivalent energy-flow description, is one of adequate smoothness of variation of relevant cochlear
properties (the mechanical and geometrical properties of the cochlear cross-
section) from base to apex. We may describe the required smoothness of variation, in energy-flow terms, as that needed to avoid passive reflexion of wave

energy flow. In a damaged cochlea, of course, such passive reflexion occurs at local irregularities or discontinuities of properties (and, in combination with re-reflexion at the base of the cochlea, may help to mediate tinnitus resonances). However, to model the mechanics of a normal cochlea, the assumption that properties of the basilar membrane and other important features of the cochlear partition vary smoothly along it, seems in accord with the available data. (For an assessment of the assumption of high frequency, see section 3.)

Firmly underlying the energy-flow description is the concept of wavenumber, and it is important to recognize that this is a precisely defined quantity, directly related to one of the quantities (the phase) that is most readily observable in measurements of basilar-membrane response to a pure tone. Of course, the key observation that first suggested a travelling-wave interpretation of cochlear mechanics was a progressive reduction in phase (that is, increasing 'phase lag') as distance from the base was increased. In that context the wavenumber, in mm^{-1}, can be defined as a rate of change: the rate at which the phase (in radians) decreases per millimetre of distance along the cochlea (Eq.(5) below). This wavenumber, k, is a measure of 'crinkliness' or 'waviness' - although it is not (in spite of its name) a sort of local 'number of waves' per millimetre; which indeed, since the phase change in a whole wave is 2π, would be $(k/2\pi)$. Similarly, in terms of the commonly used circular frequency or radian frequency ω, the frequency in hertz (cycles per second) is $(\omega/2\pi)$; but in cochlear response to a pure tone the difference between the two cases is that ω takes a constant value while the measure of 'waviness' k varies, becoming greater with distance from the base: the vibrations are sinusoidal in time but not with respect to place.

At any one place (specified by its distance x from the base, measured along the cochlear partition), there is necessarily a 'dispersion relationship' between ω and k. This identifies the value of k arising in experiments using pure tones at each different frequency ω. Conversely, for vibrations where the basilar membrane's 'waviness' takes the value k, the dispersion relationship specifies the frequency ω of pure tones for which that wavenumber k would be found at the place in question.

This latter specification is of more fundamental significance from the mechanics standpoint. In any mode of vibration of any mechanical system that is only lightly damped, the relationship

$$\omega = \left(\frac{s}{m}\right)^{\frac{1}{2}} \tag{1}$$

holds between frequency ω, stiffness s and inertia m. Here, the definitions of
s and m are such that the system's potential energy and kinetic energy are

$$\frac{1}{2} sh^2 \text{ and } \frac{1}{2} m\left(\frac{dh}{dt}\right)^2 \tag{2}$$

in terms of some measure, h, of the displacement of the system in the mode in
question. Equation (1) represents the fact that vibrational energy is, on the
average, shared equally (while being transferred back and forth) between these
potential and kinetic forms. Thus, for propagation of the primary mode of
vibration of the basilar membrane (as supported by the spiral lamina and the
bony shelf), the dispersion relationship between ω and k at any place x takes
the form of Eq. (1), where s and m represent the stiffness and inertia for
that mode in vibrations of wavenumber k. Here, the stiffness s is that of the
cochlear partition, being associated with a potential energy per millimetre
length of cochlea residing almost entirely in the basilar membrane itself; but
the kinetic energy per millimetre length resides not only in the partition but
also in the fluid, so that the inertia per unit length,

$$m = m_p + m_f , \tag{3}$$

includes both the partition inertia, m_p, and a major fluid contribution, m_f,
whose dependence upon k will be seen to be of particular importance.

High-frequency asymptotics are equivalent to the principle that, in any travelling-wave system that is only lightly damped, wave energy is propagated at a
velocity

$$U = \frac{d\omega}{dk} \quad \text{(derivative keeping x constant)}, \tag{4}$$

equal to the gradient of the dispersion-relationship curve at the place in
question. Energy flows towards the apex at a rate UE per second, where E is
the vibrational energy (potential and kinetic) per unit length of cochlea.
Here, Eqs. (1) and (2) imply that vibrations of amplitude a given by

$$h = a(x)\cos\left[\omega t + \theta(x)\right], \text{ where } \frac{d\theta}{dx} = -k, \text{ have } E = \frac{1}{2} sa^2 . \tag{5}$$

The local damping coefficient D is defined so that the rate of dissipation of vibrational energy per unit length of cochlea (due to viscous, and any other, effects) is DE. The energy loss DEdx in an interval dx defines the reduction in energy flow UE in that interval. Thus,

$$\frac{d(UE)}{dx} = -DE, \text{ giving } E(x) = \frac{U(O)E(O)}{U(x)} \exp\left[-\int_o^x \frac{Ddx}{U}\right]. \tag{6}$$

The whole energy-flow description of cochlear mechanics is summarized in Eqs. (1) to (6). They make the necessary bridge between (a) the mechanical properties assumed at each position x; namely, s, m_p, m_f and D and their dependence on k (we shall see in section 3 that the variation of m_f with k is the most important); and (b) the associated distribution of amplitude and phase in response to a pure tone of fixed frequency ω. Thus, for given ω, Eqs. (1), (2) and (3) specify the value of k for each x, and Eq. (4) that of U; whence Eq. (6) gives E, and then Eq. (5) shows how the phase of the vibrations is obtained from k and the amplitude from E.

3. CONCLUDING DISCUSSION

The form of Eq. (6) helps greatly in interpreting the most striking feature of basilar-membrane response curves: their very sharp, yet highly asymmetrical, peak, with a precipitously steep falling away beyond it. Such behaviour is assured by one very important condition: that the energy propagation velocity U falls to zero 'somewhere'; that is, at some place x (when ω is fixed) or, equivalently, at some frequency ω (when x is fixed). Then, for fixed ω, Eq. (6) with light damping D makes E(x) increase more and more as U(x) becomes less and less until, in a narrow 'critical layer' just before U becomes zero, the energy is absorbed because the integral (where dx/U represents an element of time during which the damping rate D operates) increases without limit.

The above very important 'critical-layer absorption' condition, that U falls to zero 'somewhere', is known (Lighthill 1981) to demand dispersion curves of the general character shown, for different fixed x, in Fig. 1; where, evidently, the gradient (4) tends to zero (while the wavenumber increases without limit) as the frequency ω rises to the 'resonant' value $\omega_r(x)$. Conversely, the broken line indicates how, for a tone of fixed frequency ω, the same trends occur as x increases to the value for which $\omega_r(x) = \omega$.

Actually, dispersion curves as specified by Eqs. (1) and (3) take the forms shown in Fig. 1 if m_f falls continuously from large values (much greater than m_p) to zero as k gets larger and larger, while s remains essentially constant.

68

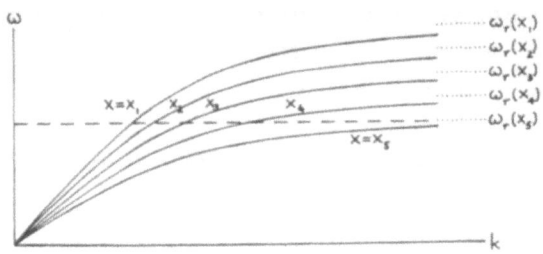

Fig. 1. Dispersion curves assuring 'critical-layer absorption'. The curve appropriate to each place x (given five values in Fig. 1) tends asymptotically to a resonant value $\omega = \omega_r(x)$ as the wavenumber k increases.

Of course the value of m_f depends, as Eq. (2) indicates, on the assumed mode of displacement of the basilar membrane and on our choice of a measure, h, of that displacement. Fortunately, hydrodynamics allows us to derive a simple form of m_f for any mode of displacement; say, a displacement

$$h\zeta(y) \text{ for } 0<y<2\ell; \text{ where } \int_0^{2\ell} \zeta(y)\,dy = 1 \tag{7}$$

may be recommended as a simple normalising condition that defines the measure h as the net change in cross-sectional area of the scala tympani due to basilar-membrane displacement. Then the ratio m_f/ρ of fluid inertia to fluid density is a nondimensional quantity

$$\frac{m_f}{\rho} = \frac{\coth k\ell}{k\ell} + 2 \sum_{n=1}^{\infty} \frac{\zeta_n^2}{(k^2\ell^2 + \tfrac{1}{4}n^2\pi^2)^{\frac{1}{2}}} \tag{8}$$

where

$$\zeta_n = \int_0^{2\ell} \zeta(y)\cos\frac{n\pi y}{2\ell}\,dy \tag{9}$$

is also nondimensional (and, by Eq. (3), $\zeta_0 = 1$). Figure 2, though calculated on just one of many possible assumptions about the basilar membrane's primary mode of bending, is qualitatively typical of the form of Eq. (6) in all cases. It shows, in particular, that 1D theory is grossly inaccurate (even though, because it meets the critical-layer absorption requirement, the errors in its predictions are reduced) and that a significant feature in the curves' shape is absent from 2D theory.

The principal approximation underlying Eq. (8) (replacing the cochlear cross-section by a square of side 2ℓ with the same area) is expected to reduce its

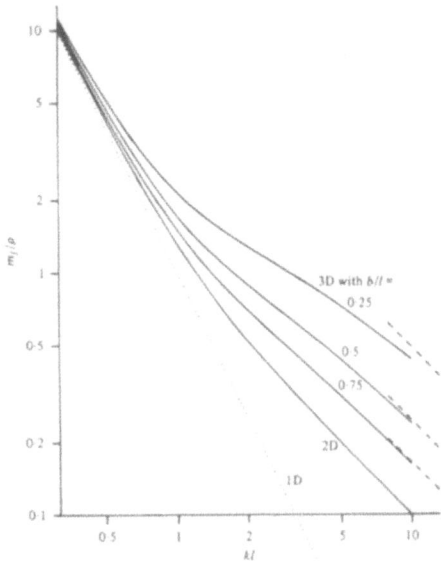

Fig. 2. Fluid inertia plotted against wavenumber, illustrated for a half-sinusoid mode of bending in a basilar membrane whose width 2b takes values 0.25, 0.5 or 0.75 times that assumed for the cochlear cross-section (Lighthill 1981). The curves marked 1D and 2D are as derived by one-dimensional theory.

accuracy very little. On the one hand, for small kℓ, Eq. (8) becomes

$$\frac{m_f}{\rho} = \frac{1}{(k\ell)^2} \; ; \tag{10}$$

and this coincides, for a cochlear cross-section of area $4\ell^2$ and arbitrary shape, with the conclusions of one-demensional theory, which are known to become correct for small kℓ (less than about 0.5). On the other hand, for large kℓ, the fluid motion becomes progressively confined to a narrow layer of thickness k^{-1} near the basilar membrane; which is both why Eq. (8) then becomes proportional to $(k\ell)^{-1}$, and why the exact shape of cochlear cross-section becomes unimportant in this limit too.

A brief comment may be added concerning the accuracy of the method of high-frequency asymptotics that utilizes these results. Comparing exact one-dimensional theory with and without simplification by the assumptions of high-frequency asymptotics, we can deduce that these produce inaccuracies only for $k < 0.2$ mm^{-1}; and we can safely use this result because Eq. (8) reduces to Eq. (10), its one-dimensional form, as soon as $k < 0.7$ mm^{-1} (Lighthill 1981). The condition $k > 0.2$ mm^{-1} for accuracy of high-frequency asymptotics is satisfied throughout the cochlea for frequencies above about 500 Hz. At lower frequencies it is satisfied except in a limited region near the base, and a simple method is available (Lighthill 1981, Fig. 14) for correcting results from high-frequency asymptotics to allow for its inaccuracy in this region.

The fluid motions whose inertia is represented by Eq. (8) are a combination of obliquely travelling wave-like modes in each of which the fluid particles describe circular paths, and a longitudinal mode where the same is true for $k\ell > 1.5$ (that is, as resonance is approached). They are potential flows outside a 'Stokes boundary layer' of displacement thickness $(\nu/\omega)^{\frac{1}{2}}$, characteristic of all oscillatory motions, where the kinematic viscosity ν of the cochlear fluids at body temperature is about $0.7 \text{ mm}^2 \text{ s}^{-1}$; this thickness, about 0.01 mm at 1 kHz, is too small to influence the fluid inertia. The boundary layer gives, however, a viscous-dissipation contribution

$$\rho s^{-1} \omega^2 (\tfrac{1}{2}\nu\omega)^{\frac{1}{2}} \tag{11}$$

to the damping rate D, which (if it is the main contribution) is consistent with the assumption $(D \ll \omega)$ of only light damping.

We conclude with comments on m_p and s. The inertia of the cochlear partition, m_p, is independent of k; but, as with m_f, the value to be ascribed to it depends on the mode (and on the measure) of basilar-membrane displacement. Rewriting Eq. (C42) of Lighthill (1981) in the notation of this paper, the assumptions in Eq. (7) give

$$m_p = \int_0^{2\ell} M(y) \left[\zeta(y)\right]^2 dy, \tag{12}$$

where M (y) is the mass per unit area of the basilar membrane and of all the other solid structures that move with it.

Finally, the subject's major paradox needs to be highlighted once again. It is that the dispersion curves given by Eqs. (1) and (3), subject to behaviour of m_f and m_p as outlined above, will take the 'critical-layer absorption' form (Fig. 1) provided only that the stiffness s is essentially independent of k. This demands (Lighthill 1981, p. 168) that transverse stiffness dominates over longitudinal stiffness in vivo. Such a paradoxical conclusion appeared quite incompatible with the measured properties (similar to those of 'an elastic plate') of basilar membranes taken from cadavers, until Voldrich (1978) discovered experimentally that the longitudinal stiffness was indeed negligible in vivo although quickly becoming significant after death.

REFERENCES

De Boer, E. (1979) Short-wave world revisited: resonance in a two-dimensional cochlear model. *Hearing Research 1,* 253-281.
Holmes, M.H. (1982) Dynamics of the inner ear. *Journal of Fluid Mechanics 116,* 59-75.
Lighthill, J. (1978) *Waves in Fluids* (Cambridge University Press, Cambridge).
Lighthill, J. (1981) Energy flow in the cochlea. *Journal of Fluid Mechanics 106,* 149-213.
Steele, C.R. (1974) Behaviour of the basilar membrane with pure-tone excitation. *Journal of Acoustical Society of America 55,* 148-162.
Steele, C.R. and Taber, L.A. (1979) Comparison of 'WKB' and finite difference calculations for a two-dimensional cochlear model. *Journal of Acoustical Society of America 65,* 1001-1006.
Viergever, M.(1980) *Mechanics of the Inner Ear* (Delft University Press, Delft).
Voldrich, L. (1978) Mechanical properties of basilar membrane. *Acta Otolaryngologica 86,* 331-335.
Whitham, G.B. (1974) *Linear and Nonlinear Waves* (Wiley, New York).

Section III

Cochlear emissions

AN INTEGRATED VIEW OF COCHLEAR MECHANICAL NONLINEARITIES
OBSERVABLE FROM THE EAR CANAL

D.T. Kemp, A.M. Brown

Institute of Laryngology and Otology
Gray's Inn Road, London WC1X 8EE

ABSTRACT

Otoacoustic emission phase has been examined as a function of stimulus frequency. Two components have been compared. The emission of the acoustic product 2f1-f2 has phase characteristics which are broadly derivable from current nonlinear cochlear models. An explanation of stimulus frequency re-emission phase requires the postulation of multiple fixed place reflection/retransmission sites along the spiral organ. It is shown how the distortion product emission phenomenon can be used as an experimental probe to determine the sites of stimulus frequency retransmission.

1. INTRODUCTION

Detailed analysis of the sound pressure in the sealed ear canal during acoustic stimulation reveals a small complex nonlinear component to be present at all stimulus levels (Kemp and Brown 1983). This component has been shown to be vulnerable to noise exposure, ototoxic drugs and physiological disturbances in the same way as is cochlear function (Anderson and Kemp 1979, Kemp 1982). Many features of the two tone interactions exhibited by this nonlinearity closely resemble those found in the cochlea (Kemp and Chum 1980, Brown and Kemp 1983). It is widely accepted that this nonlinear contribution to ear canal sound pressure is due to nonlinear biomechanical activity in the cochlea.

There are two types of stimulated otoacoustic emissions with quite different intensity growth functions (Kemp and Brown 1983). The first type comprises the return of energy at the stimulus frequency/frequencies to the ear canal, i.e. stimulus frequency emissions or SFE's. Continuous acoustic excitation of the cochlea results in continuous stimulus frequency otoacoustic emissions as described by Kemp and Chum 1980. The SFE level progressively saturates with stimulus level increase.

The second type of emission arises from intermodulation within the cochlea when two stimulus tones are presented. Notably the distortion product 2f1-f2 is emitted (Kemp 1978, Kim 1980, Kemp and Brown 1983, Brown and Kemp 1983). We refer to this as DPE or distortion product emission. Its level tends to be a constant proportion of stimulus level.

Techniques for observing SFE and DPE phenomena are described in the literature cited above. For frequency domain studies these generally involve phase locked narrow band filtering of the ear canal sound pressure signal to select the frequency component required, and often use suppression by an additional tone to identify the physiological element.

The aim of this paper is to explore the relationship between SFE's, DPE's and cochlear wave propagation, and to take a first step towards modelling the unique SFE signal generated by individual ears.

2. ACOUSTIC DISTORTION PRODUCT GENERATION BY THE COCHLEA

The existence of DPE's can be predicted from nonlinear basilar membrane theory. The propagation of DP energy back from the site of generation to the base of the cochlea was modelled by Hall in 1974, but the conclusion that DP sound pressure would appear in the meatus was overlooked for some years. Although DPE does have a theoretical foundation the detailed comparison of experimental DPE data with measured cochlear parameters and with nonlinear model predictions is in its early stages.

We can predict certain phase characteristics of the DPE from cochlear wave propagation data. A specific example for cat is presented and discussed in Fig. 1. With reference to Fig. 1, we consider below the expected phase behaviour of the DPE under two conditions of stimulus frequency change. For this purpose we presume that the vectorial spacial sum of the DP produced in the intermodulation area between the f1 and f2 places constitute a single source of DP with amplitude and phase ϕdp monotonically dependent on f1 and f2. We take the (resultant) source of DP to be a place on the basilar membrane from which DP energy is effectively transmitted both apically to the DP resonant place and basally to create the DPE.

At the place of generation the phase of the distortion product with respect to that derived directly from the stimuli, is given by $2\phi f1 - \phi f2$ where $\phi f1$ and $\phi f2$ are the stimuli phases at the interaction place. At the ear canal there is a small additional lag of ϕfdp due to reverse propagation. Referring to Fig. 1, if both f1 and f2 are changed by the same proportion (i.e. keeping a fixed spacial separation on the BM) no phase changes of f1 or f2 occur at the DP generation site. The phase of DPE should therefore, remain constant during iso-ratio stimulus sweeps, dependent of course upon the true logarithmic frequency characteristics of basilar membrane mechanics. The DPE would appear to have zero group latency although the true phase delay would be

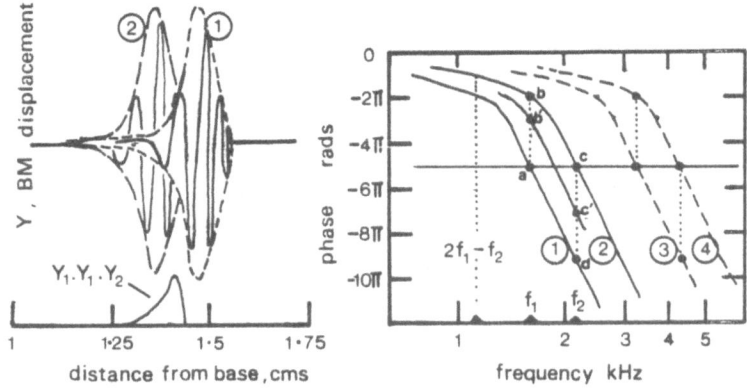

Fig. 1. Propogation along the basilar membrane and DP generation. Extrapolated neural data for cat after Neely and Kim 1983. Left: This shows a derived basilar membrane vibration pattern during two tone stimulation at 2.2 kHz (2) and 1.65 kHz (1) respectively. In a simple asymmetrically nonlinear system the level of the intermodulation component 2f1-f2 generated would depend upon the square of the f1 amplitude, Y1 times the amplitude of f2, Y2. This function, performed on the actual vibration envelopes shown is also given. The median is between the f1 and f2 peaks but nearest to the f2 (higher frequency) peak. With other stimulus frequency ratios the intermodulation region would remain strongly constrained by the rapid apical cut-off in f2 excitation, and would always be near to the f2 place. Right: This shows the phase characteristics of several points on the BM in our example. Lines 1 and 2 correspond to points at 1.33 and 1.44 cms, having fc's of 2.2 and 1.65 kHz respectively, matching the excitation pattern on the left. Lines 3 and 4 relate to points with fc's twice those of 1 and 2. We have generated this data by laterally shifting 1 over the log frequency scale. Resonance in this idealised cochlea occurs at a phase lag of 5 π at each points, marked by the horizontal line. The point (b) gives the phase of f1 at the f2 place, and point (a) gives the phase of f2 at the f1 place. Since maximum DP production occurs between f1 and f2 places, we take points b' and c' to be a better guide to the actual phase lags of the two stimuli at intermodulation. These are $\phi f1=3\pi$ and $\phi f2=7\pi$ respectively. The lag for the DP frequency (2f1-f2), ϕdp is seen to be small (-0.75π). We presume that this lag applies also to reverse transmission so that the phase of DPE is given by $\phi dpe=2\phi f1-\phi f2+\phi dp$.

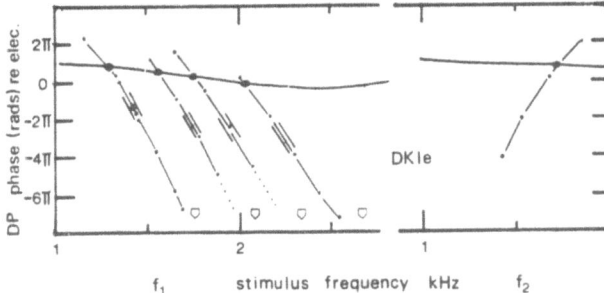

Fig. 2. The phase of human DPE under different stimulus frequency conditions. Stimuli were f1,75 dB SPL and f2, 70 dB SPL. The solid line is for iso-ratio stimulation with f1/f2=0.75. Left: Dashed lines are for fixed f2, swept f1 stimulation. Each line is for a different f2 value, marked by the arrow. Intersection with the solid line occurs when f1/f2=0.75. The triple line section marks where the maximum DPE amplitude was found. This occurred around f1/f2=0.8. The DPE was too small to obtain phase data at the ends of each line. The choice of f2 was not critical. Results were confirmed on 5 other ears. Right:fixed f1 and swept f2, the phase changes are reversed but not halved as expected from $\phi dp=2\phi f1-\phi f2$.

many waves.

The zero group latency proposition was tested experimentally. Fig. 2 (solid
line) shows DPE phase data for a human ear tested at f1/f2=0.75. There is only
a modest phase change over the 2 kHz stimulus frequency sweep. This result has
been confirmed on 14 other ears using f1/f2=0.8. In Fig. 3, DPE amplitude and
phase data for three human ears is given with an expanded phase scale. Each
ear shows a similar $\pm \pi$ change over the 2.2 octave range with maximum lag at
3.5 kHz. This limited frequency dependence might point to a systematic depart-
ure from logarithmic performance in the human ear but is more likely to be due
to the transmission characteristics of the middle ear. The implied DPE latency
is less than ½ millisecond. Despite the flat phase characteristics of the iso-
ratio DP measurements, the underlying mechanism can be shown to be sharply
tuned by suppression experiments (Brown and Kemp 1983 Kemp and Brown 1983).

Fig. 3. Iso-ratio measure-
ments of the DPE amplitude
and phase for 3 human ears
from 1 to 5 kHz. Stimuli
were f1, 75 dB SPL, f2, 70
dB SPL, and f1/f2=0.75. The
transducer amplitude freq-
uency responses have been
digitally subtracted from
the data. Artefactual dis-
tortions were below the
instrumental noise floor.

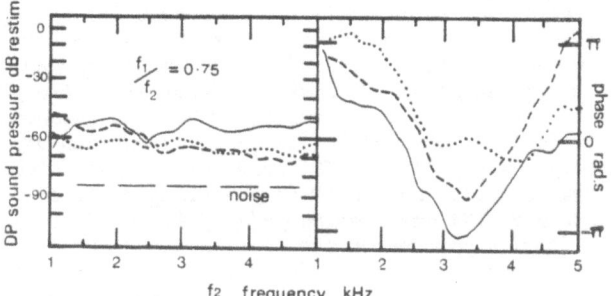

The DPE was suppressible by 10 dB by an 80 dB tone at 0.9 f2, but not by a
similar tone at 1.3 f2, i.e. the DPE was tuned.

We next consider what would happen if only stimulus 1 is increased in frequency. The lag of f1 will increase over the f2 excitation region. If the place
of interaction remains relatively fixed near f2, and ϕdp is small then $d\phi dpe=$
$2d\phi f1$. Thus the DPE acts as a 'carrier' of f1 phase changes at the interaction
place. The rate of change of DPE phase with f1 frequency directly depends upon
the group latency of f1 at the interaction site. In the cat example (Fig. 1)
with f2=2.2 kHz, the group latency of f1 at the estimated interaction site is
given by the slope at b' and is 2.8 milliseconds, or approximately 3 DP freq-
unecy waves.

Experimental data for gerbil tested under the above conditions (Kemp and Brown
1983) conforms to these predictions. The DPE latency was found to be just over
2 milliseconds. In human ears greater latencies are found. Fig. 2 (dashed
lines) gives examples of human DPE phase change under fixed f2, swept f1
conditions. Phase changes of around 8π are seen over the 1/3 octave f1 range
for which significant interaction is observed. These data indicate a 6 wave

latency of fl at the DP interaction place.

3. STIMULUS FREQUENCY RE-EMISSION BY THE COCHLEA

The re-emission of stimulus frequency energy by the cochlea is not an inherent property of current cochlear models. Nevertheless the phenomenon occurs to a substantial degree at low stimulus levels. For human subjects the SFE is typically 10 db SPL during 40 dB stimulation as shown for two human ears in Fig. 4. We must accept that the signal carried by the travelling wave can be retransmitted or reflected by some biophysical mechanism.

If we were to postulate that the retransmission mechanism was an integral part of the physiological response sharpening process then it would be active over some specific part of any excitation pattern; say at the peak. By the same reasoning we applied to iso-ratio stimulation of DPE's, we would expect zero group latency for the SFE.

A low-latency SFE has recently been found in gerbil at moderate to high stimulus levels (Kemp and Brown 1983). However at low levels in gerbil and at all stimulus levels in man the dominant nonlinear SFE component has a high group latency. Fig. 5 shows the rapid increase in SFE phase lag with stimulus frequency for human ears. The group latency is of the order of 10 waves.

If Figs. 5 and 2 are compared in detail we find that SFE phase gradients match DPE phase gradients when fl is swept towards a fixed f2. This suggests a model for delayed SFE generation.

In DPE generation the stimulus frequency f2 largely defines the point on the BM at which the DP is created. As discussed in section 2, DP transmissions from this place 'carry' the phase lag of fl at that place. If there were localised regions of the BM from which stimulus retransmission of fl occurred, then we would expect just the phase agreement observed. The fl phase lag doubling inherent in 2fl-f2 generation would be matched in the SFE phenomenon by the doubling of phase lag in the reverse transmission of fl. Both the SFE and the DPE would have the same phase gradients with fl, provided the f2 place coincided with the SFE transmitter point.

The range of frequencies over which substantial stimulus retransmission might be obtained from a single small region of anomolous biomechanical activity would depend on the width of the excitation pattern. By the DPE-SFE analogy

above this range can be estimated from Fig. 2 to be 1/3 to 1/2 octave so that
several sources would be needed to explain the frequency extent of SFE shown
in Fig. 4. In Fig. 4c we show the result of an attempt to model just one section of the SFE in 4b with an artificial transmitter point defined by an f2 of
2.15 kHz. The phase of f1 obtained via the DP process matches the phase of f1
via the SFE process over a short frequency range. Clearly by an iterative process the SFE data pattern could be transformed into a set of f2 frequencies of
various amplitudes. It remains to be seen whether a unique solution exists for
a particular ear and whether the process would actually identify significant
sites in the cochlea.

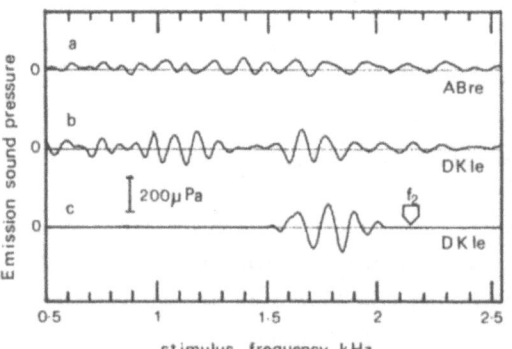

*Fig. 4 (a) and (b): Recordings of
stimulus frequency acoustic emissions from two human ears. Only the
in phase amplitude is shown. The
quadrature phase component looks the
same but for a 90 degree shift in
the amplitude oscillations with
frequency. The oscillations signify
increasing SFE phase lag with frequency with the SFE being alternat-
ely in phase (+ve) and out off
phase (-ve) with the stimulus. Part
(c): a sample recording of DPE during an f1 sweep, with f2 fixed.
Note the DP phase and amplitude
envelope roughly matches the SFE
over a small range.*

*Fig. 5. SFE phase data extracted
from Fig. 4 (a) and (b). At
frequencies of very low output and
where latency changed abruptly the
trace has been broken and continued
at zero lag.*

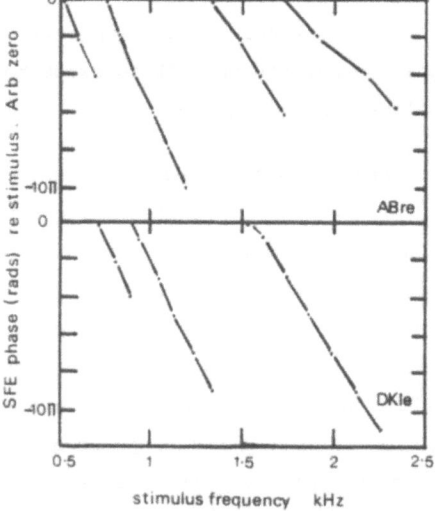

So far we have treated the DPE and
SFE phenomena as if occurring in
isolation from each other. In fact,
any intracochlear DP signal will
act as a stimulus and excite the
SFE mechanism. The result is a fine
structure of the iso-ratio DPE. In
Fig. 6 we show iso-ratio DPE data with fine structure clearly due to the SFE
mechanism. The fine structure 6 (a) is selectively suppressed by low level
tones near to the DP frequency 6 (b). This allows isolation of the fine

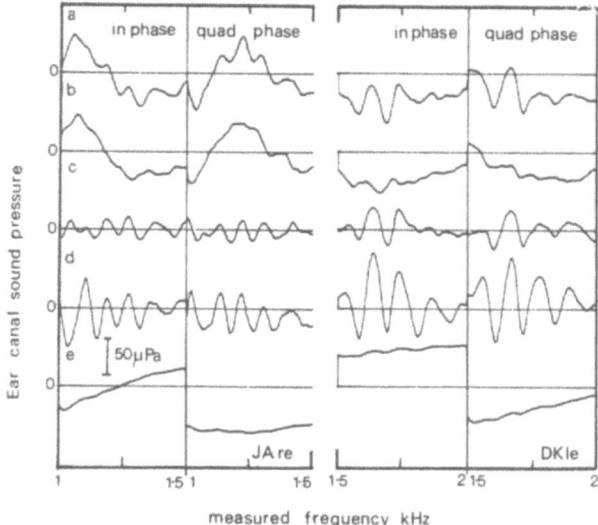

measured frequency kHz

Fig. 6. Two component (in phase and quad phase) recordings of acoustic emission components for two ears each over different small frequency ranges. Abscissa refers to the frequency of the component displayed, not the stimuli. The fine structure is of interest. (a): Raw DP data iso-ratio sweeps with f1/f2=0.75 and levels 70 and 75 dB SPL. Note the ripples on the slowly changing DP component. These ripples were selectively suppressible by a tone of 1.1 fdp at 50 dB SPL and above. This allows separation of two DP components. The robust component is (b) and the easily suppressible one (c) shows considerable latency. The SFE for a 40 dB SPL stimulus at the DP frequency is shown in (d). This matches well the easily suppressible DP component (c). Clearly (c) is the SFE mechanism responding to the DP acting as a secondary stimulus within the cochlea. Finally (e) is the f1 stimulus in the meatus reduced by a factor of 30 to fit the otherwise fixed scaling. The ripples in the stimulus levels are due to interference by the SFE as are the ripples in the DP level (a), albeit at a second order level.

structure component 6 (c), which then matches the SFE obtained for an externally applied stimulus of frequency DP, 6 (d).

4. SUMMARY AND CONCLUSIONS

We have shown that the acoustic distortion product 2f1-f2 can be evoked by any close pair of tones and its phase behaviour can be predicted. The 2f1-f2 emission does not seem to depend upon intrinsic inhomogeneities in cochlear mechanics, indeed the iso-ratio data confirms a high degree of homogeneity. The only localised factor is that created by the f2 excitation itself. In contrast, stimulus frequency emissions clearly do need intrinsic localised anomalies to explain their phase properties. With selected f2 values, DP emission phase can match SF emission phase very well over a 1/3 octave frequency range below f2. We propose that this particular f2 defines an SFE-active site in the cochlea. Previous work (Kemp and Chum 1980) has shown the SFE to be most easily suppressed by slightly higher frequency tones. The saturating property of the SFE in contrast to the DPE perhaps implies that the mechanical anomaly is smoothed out at higher stimulations.

82

We find no basis for attributing signal delays to the SFE mechanism over and above those normally present in the travelling wave, once the doubling effect is accounted for. We do find latency in human ears to be greater than that found in laboratory animals under the same stimulus conditions between 1 and 4 kHz. We propose the DPE phenomenon as a probe with which to explore cochlear wave propagation times, and the SFE phenomenon.

Acknowledgements

This work has received support from the Medical Research Council and the Iron Trades Insurance Group.

REFERENCES

Anderson, S.D. and Kemp, D.T. (1979) The evoked cochlear mechanical response in laboratory primates. *Arch. Otorhinolaryngol.* 224, 47-54.
Brown, A.M. and Kemp, D.T. (1983) Suppressibility of the 2f1-f2 stimulated acoustic emissions in gerbil and man. *Hearing Research* (submitted).
Hall, J.L. (1974) Two-tone distortion products in a nonlinear model of the basilar membrane. *J. Acoust. Soc. Am.* 56, 1818-1828.
Kemp, D.T. (1979) Evidence of mechanical nonlinearity and frequency selective wave amplification in the cochlea. *Arch. Oto-Rhino-Laryngol.* 224, 37-45.
Kemp, D.T. (1982) Cochlear echoes-implications for noise-induced hearing loss. In:*New perspectives in noise-induced hearing loss* (D. Henderson *et al.*, eds). pp 189-207, Raven Press, New York.
Kemp, D.T. and Chum, R. (1980) Observations on the generator mechanism of stimulus frequency acoustic emissions two-tone suppression. In: *Psychophysical, physiological and behavioural studies in hearing.* (van den Brink, G., Bilsen, F.A. eds). pp 34-41. Delft University Press.
Kemp, D.T. amd Brown, A.M. (1983) A comparison of mechanical nonlinearities in the cochleae of man and gerbil from ear canal measurements, In: *Hearing Physiological Basis and Psychophysics.* (R. Klinke and R. Hartmann, eds) In Press. Springer Verlag, West Germany.
Kim, D.O. (1980) Cochlear mechanics: implications of electrophysiological and acoustical observations. *Hearing Res.* 2, 297-317.
Neely, S.T. and Kim, D.O. (1983) An active cochlear model showing sharp tuning and high sensitivity. *Hearing Research* 9, 123-130.

MODELLING COCHLEAR ECHOES: THE INFLUENCE OF IRREGULARITIES IN FREQUENCY MAPPING ON SUMMED COCHLEAR ACTIVITY

G.J. Sutton, J.P. Wilson

Department of Communication and Neuroscience, University of Keele, Staffs., ST5 5BG, U.K.

ABSTRACT

"Cochlear echoes" occur in restricted frequency bands specific to each ear, with a long peak latency. A computer simulation has been developed based on basilar membrane vibration amplitude and phase followed by a "second filter" process. The vector sum of the second filter outputs, representing gross cochlear activity, is very small for a smoothly changing basilar membrane impedance and regular tonotopic mapping of the second filters. However, any small irregularity in either of these functions leads to a greatly enhanced response in the corresponding frequency region, which simulates the bandwidth, latency, and other features of cochlear echoes.

1. INTRODUCTION

Two types of model have been proposed to explain cochlear echoes. Kemp (1978) and Kim *et al.* (1980a) have suggested a reverse travelling wave, reflected from an impedance discontinuity on the basilar membrane (BM), to explain the observed long latency and frequency specificity. Wilson (1980bc) suggested an alternative model, based on an irregularity in frequency mapping and the synchronous swelling and shrinking of hair cells following a second filter (SF) process, in order to explain an experimentally observed zero reverse travel time and limited maximal amplitude. This model was designed to give minimal local influence on BM mechanics, but recent direct measurements, showing sharp basilar membrane tuning (Khanna and Leonard, 1981, and Sellick *et al.*, 1982) now appear to require local feedback to occur. Furthermore, the high levels of ear-canal distortion product found by Kim *et al.* (1980b) and the high levels of spontaneous oto-acoustic emission reported by Glanville *et al.* (1971) and followed up by Wilson and Sutton (1983), render the hair cell swelling model less plausible. Nevertheless, other features of the model have been notably successful in simulating cochlear echoes whereas a specific test based on a reverse travelling wave model (Neely, 1981) was not. The present paper develops the consequences of such a two-stage model, in which gross cochlear activity is unspecified. Active and non-linear processes do not feature in the model but could easily be incorporated.

2. METHODS

A one-dimensional model of the BM was utilised based on a program of de Boer (1980), after Allen (1977). The parameter values were chosen to be approximately representative of values for man. The impedance of the cochlear partition is given by:

$$Z(x) = (C_0/i\omega)\exp(-\alpha x) + i\omega M_0 + R_0\exp(-\alpha x/2) \tag{1}$$

with stiffness, $C_0 = 10^9$ dyn. cm^{-3}; mass, $M_0 = 0.05$ g cm^{-2}; resistance, $R_0 = \delta (M_0.C_0)^{\frac{1}{2}}$; with damping, δ, either 0.05 or 0.20; and $\alpha = 3$ cm^{-1}; length of cochlea = 3.5 cm; height of scalae, h=0.1 cm. This gives a "true" resonance at the point where stiffness and mass components cancel. The resistive term gives a loss factor δ constant with respect to x, which makes the travelling wave pattern invariant as a function of frequency or position (mapping coefficient, 4.62 mm/octave). Two values of δ (= 0.05, 0.20) were used in order to model both highly resonant and "classically" damped types of response. The calculated frequency response of one point (at which the stiffness and mass components of Z(x) cancel for 1 kHz) for both the values of δ are illustrated in Fig. 1 (BM).

A simple resonant second-order "second" filter (SF) was then incorporated with a Q value adjusted to give a combined filter bandwidth of about 140 Hz at 1 kHz (i.e. a Q_{10dB} of about 7). The values of Q required were 8 and 20 for δ = 0.05 and 0.20 respectively. The SF resonance frequency was arbitrarily made equal to that of the BM for the low damping case (3.6dB down the HF cut-off) and 3dB down (i.e. below resonance in this case) for high BM damping. This procedure gives the responses shown in Fig. 1 (BM+SF); with the SF response normalised to give unity gain at its centre frequency for Q=8.

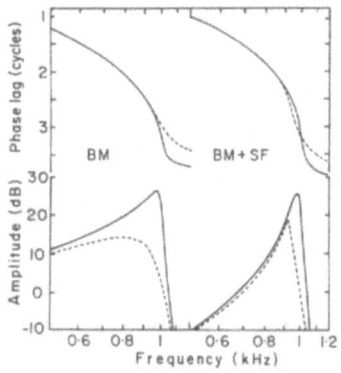

Fig. 1. Phase and amplitude responses for cochlear model, relative to stapes motion. Left: for a point on basilar membrane (BM) with damping factors (δ) of 0.05 (solid line) and 0.20 (dashed). Right: after 'second filter' (BM+SF), centred at 1001 Hz with Q=8 for low damping, and at 929 Hz with Q=20 for high damping.

For a given input frequency, the program calculated at each point the amplitude and phase of the BM response (represented by a vector in the complex plane) after which the filtering effect of the SF at each point is incorporated, thus giving a vector which represents the amplitude and phase of the activity at that point. This is shown diagrammatically in Fig. 2c. This can be imagined as rotating about the x-axis in time, and Figs. 2b and 2d are then

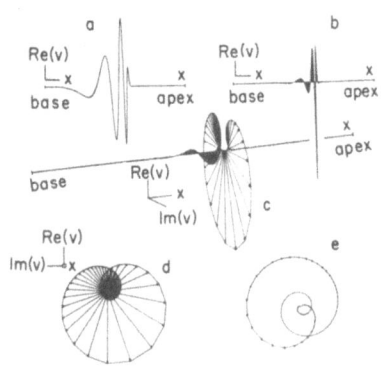

Fig. 2. Waveform as a function of distance (x) along the BM for a 1 kHz input. (δ = 0.20, Q=20). Re(v) and Im(v) refer to the real and imaginary parts of the (complex) excitation. (a) BM vibration pattern, (b) Effective excitation pattern after SF. (Net area under the curve is proportional to the instantaneous SR.) (c) Representation of (b) in complex vectors. The wave pattern 'rotates' in time about the x-axis. (d) Projection of (c), looking down the x-axis. (e) Calculation of summed response (SR) by laying vectors end-to-end to form a phasor diagram. Resultant SR is distance between ends of chain (small with no irregularity).

2-dimensional projections of it. These vectors are added to give a resultant 'summed response' (SR) representing the net mechanical response or unweighted cochlear microphonic potential. This is illustrated in Fig. 2e where the vectors of Fig. 2d are layed end-to-end to give a phasor diagram; the distance between the ends of this curve being the magnitude of the SR. It was verified that for the number of elements chosen (1000), the phase and amplitude changes between adjacent points were small enough to ensure the vector summation was sufficiently accurate (taking 4000 points made little difference).

3. RESULTS

With a 'normal' BM and logarithmic frequency to linear place mapping of the SFs the SR was found to be extremely small and approximately uniform with frequency. With any variation in BM phase and/or amplitude, however, the SR was no longer small. A small (8 element) irregularity was then incorporated into the SF mapping, such that each element had its SF centred at 1003 Hz (for δ = 0.05, 925 Hz for δ = 0.20) instead of decreasing monotonically from 1022 Hz to 985 Hz as in the regular mapping (see Fig. 3). The effect that this has on

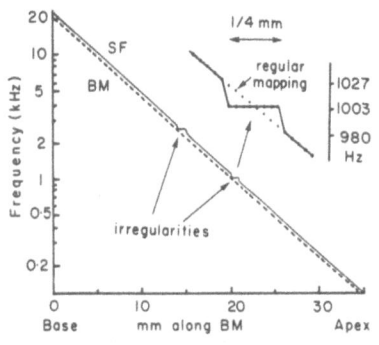

Fig. 3. Mapping of SF tuning (solid line) on to distance along BM, showing two irregularities (enlarged on inset; 8-elements ≡ ¼ mm). BM peak tuning values are shown (dashed) with an exaggerated basalward shift relative to resonance frequency.

the SR can be seen in Fig. 4, where "0 dB" represents the SR for the uniform map. In both cases there is greatly enhanced response (+ 34dB) in a narrow frequency band (50 Hz for δ = 0.05, and 30 Hz for δ = 0.2). These

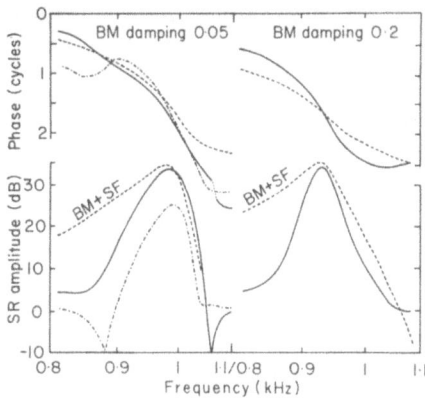

Fig. 4. Phase and amplitude of SR as a function of frequency for (left) δ = 0.05, Q=8; (right): δ = 0.20, Q=20. Solid curves: 8-element plateau in SF mapping; dot-dash: 8-element plateau in BM impedance; BM+SF curves shown for comparison. Zero amplitude is for regular SR mapping; arbitrary phase zero.

are somewhat sharper than the BM+SF curve, (dashed line), and the phase slope is steep and exceeds that for BM+SF. The 'notch' in the low damping case arises due to cancellation between the local response and the non-zero response from the regular mapping.

The impulse responses were calculated (assuming linearity) from the amplitude and phase data for BM, SF, BM+SF, and for SR: the last case representing the click-evoked echo. These calculated impulse responses for both damping values are shown in Fig. 5. For SR, we have a slow build up giving a very long delay to peak response, of about 14 cycles in both cases, compared with about 3 cycles for the BM responses and about 8 or 6 cycles for BM+SF (δ = 0.05). The picture is similar for δ = 0.20.

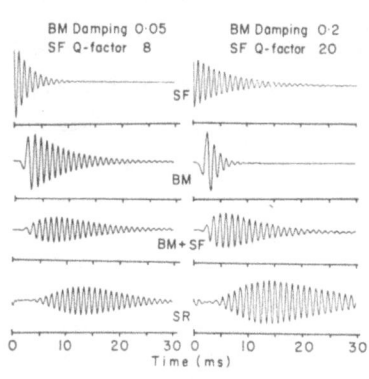

Fig. 5. Impulse responses for: Left, top to bottom: SF alone (Q=8), BM alone (δ = 0.05), combined effects of BM+SF at same point, SR over all points with 8-element irregularity in SF mapping, centred at same point. Right: as left, for Q=20 and δ = 0.20.

The effect of altering the length of the plateau region from 3 to 15 elements is shown in Fig. 6. The main effect is a change in bandwidth of the SR, with no effect on the phase slope. The impulse response for the 15-element case illustrated, shows a slightly lower latency than the 8-element case (Fig. 5, SR, left). The sharp-cornered, horizontal 'plateau' used above is somewhat unrealistic, so a smoothed curve and less steep slopes were tried successively. These merely reduced the magnitude of the SR slightly, but not its bandwidth or phase slope.

The interaction of two irregularities was also investigated. Fig. 7 shows the pattern of results found as a function of frequency, with separation of

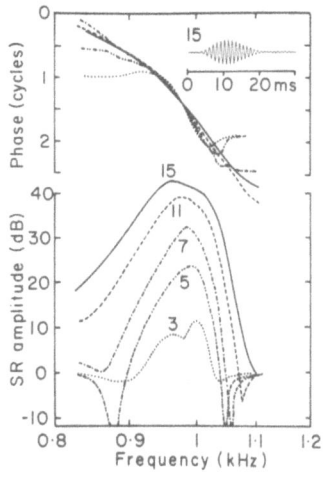

Fig. 6. SR phase and amplitude for various lengths of irregularity (3, 5, 7, 11, 15 elements). (δ = 0.05, SF Q = 8). Impulse response is for 15 element irregularity.

the two (8 element) irregularities as parameter. Below 15 elements, the patterns merged but above that two separate regions were found.

Instead of an irregularity in the mapping of the second filters, an irregularity in the cochlear partition impedance was introduced. An 8 element section of BM was given a constant impedance, corresponding to that previously occurring at the centre of that region, rather than having stiffness and resistance decreasing exponentially (see Eq. (1)). The pattern of amplitude and phase irregularity seen suggested some reflection (coefficient ∿0.1) from the impedance discontinuity. The SR

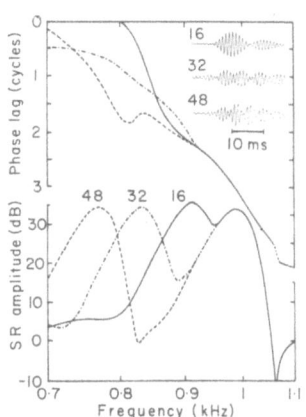

Fig. 7. SR phase and amplitude, and impulse responses, for a pair of 8-element irregularities at separations of 16, 32, and 48 elements between centres. (δ = 0.05, Q=8).

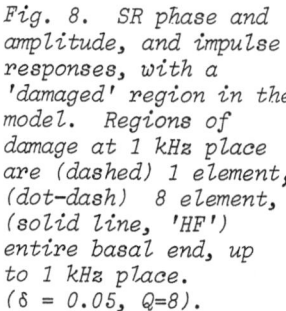

Fig. 8. SR phase and amplitude, and impulse responses, with a 'damaged' region in the model. Regions of damage at 1 kHz place are (dashed) 1 element, (dot-dash) 8 element, (solid line, 'HF') entire basal end, up to 1 kHz place. (δ = 0.05, Q=8).

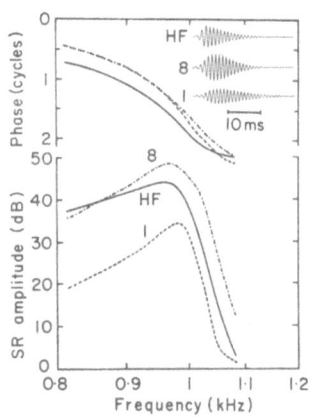

is illustrated by the dot-dashed line in Fig. 4, and is comparable to that for the SF mapping irregularity, but the phase slope is steeper (equivalent to about 20 cycles delay at maximum), and shows a reversal in one frequency region.

An alternative type of irregularity is in the sensitivity of individual elements: this is particularly relevant to localised hearing loss. It is found that even with just one element not contributing (equivalent to about 30

damaged hair cells) the expected SR is significant, and the frequency response
found (Fig. 8 dashed line) is practically the same as the BM+SF response.
Widening the 'damaged' region to 8 elements (dot-dash line) gives a broader
bandwidth, and in the extreme case where all the basal contribution is eliminated, one finds a response (solid line, HF) like BM. The corresponding impulse responses are also illustrated, and show a decreasing latency as the 'damaged' region is widened (bottom to top). These results imply that the SR is
susceptible to any non-uniformity in the sensitivity of the hair cells.

The general pattern of findings was also found to be valid for a massless BM
model, although with the parameters used the resulting amplitude and latency
were less than the model with mass and a true resonance.

4. SIMPLIFIED TREATMENT

Since the SR with a uniform BM and SF mapping is very small, one can say that
the non-uniform SR is very nearly equal to the *difference* between the response
with an irregularity and that with a uniform map. One then need only consider
the region of irregularity itself, since the contributions from the rest of
the BM will be identical in both cases.

Next, one may take the amplitude of excitation to be constant and the phase to
be linear over the region of a small irregularity. Then, since the phase of
the resultant vector over the region is the same for both regular and irregular mappings, and as the phase of the central element, the SR must have the
same phase as this element; and the phase of the SR must vary with frequency
as does the phase of the central element, i.e. as the phase of BM+SF. Under
the above assumptions the magnitude of the SR will to a first approximation
be proportional to A $\Delta\phi(\Delta\psi + \Delta\psi /2)$ where $\Delta\phi$ and $\Delta\psi$ are respectively the
change in the SF and BM phase over the irregularity and A is the amplitude of
the central vector. Agreement with the results of the full computation is
reasonably good considering the approximations made. The additional phase
slope for SR found in the comprehensive model arises from the slight change
in amplitude across the irregularity.

We conclude that the long latency which is seen in the SR is the result partly
from the inherently large phase slope of the cochlear filter and partly from
this additional factor.

5. DISCUSSION

A two-stage filter model with regular tonotopic mapping gives very little

SR due to the high degree of phase cancellation. This is specifically not so for either the BM alone (where SR must represent stapes volume displacement) or an array of SF's alone. A slight irregularity in either or both of these functions, however, leads to a greatly enhanced summed response specific in frequency to the region of the irregularity. For model values chosen to match critical bands, the calculated bandwidth and latency values turn out to correspond with those for the cochlear echo (without recourse to reverse travel time). The excess latency above that due to BM+SF can be viewed as a cancellation phenomenon attenuating the earlier part of the response. An important function of phase delay along the BM may be to give a high degree of remote cancellation of mechanical and electrical responses to maintain stability with an active system.

A mapping or sensitivity irregularity may be an early symptom of pathology. This might account for the occurrence of tinnitus after overstimulation (Kemp, 1981, Zurek and Clark, 1981) and near the edges of a region of hearing loss (Wilson and Sutton, 1981). It is possible that under strongly driven conditions the degree of sensitivity irregularity between neighbouring elements may decrease. This would then reduce the relative strength of the SR and account for the low maximum levels of cochlear echo without requiring the underlying active process to saturate at this level. Clearly the latter would be an embarrassment for a sharpening mechanism intended to work over a wide dynamic range.

The model has not considered the effect of feedback of SR to the middle ear and reflection back into the cochlea. This would be required in a full treatment to explain threshold minima, loudness enhancements, the occurrence of spontaneous emissions, etc. These effects would occur where the SR amplitude is appreciable and in phase with stapes vibration. The frequency spacing predicted by the model would then be about 70 Hz at 1 kHz, in agreement with experimental values (Kemp, 1979, Wilson, 1980a).

It appears that the active process (i.e. the generators of the SR) must have some effect in sharpening the BM response (see Introduction). Since anything that affects BM displacement must also (with rigid walls and incompressible fluid) equally affect stapes motion, this local effect is transmitted to the stapes. With stapes volume velocity specified, this will be seen as a pressure change.

The long straight phase slopes found experimentally (Wilson, 1980a) were not

90

found in the model. This may result from the limitations of the BM and SF
models used.

REFERENCES

Allen, J.B. (1977). Cochlear micromechanics - a mechanism for transforming
    mechanical to neural tuning within the cochlea. *J. Acoust. Soc. Am.* 62,
    930-939.
De Boer, E. (1980). Auditory Physics: Physical Principles in Hearing Theory
    I. *Physics Reports* 62(2), 87-174.
Glanville, J.D., Coles, R.R.A., Sullivan, B.M. (1971). A family with high-
    tonal objective tinnitus. *J. Laryngol. Otol.* 85, 1-10.
Kemp, D.T. (1978). Stimulated acoustic emissions from within the human
    auditory system. *J. Acoust. Soc. Am.* 64, 1386-1391.
Kemp, D.T. (1979). The evoked cochlear mechanical response and the auditory
    microstructure - evidence for a new element in cochlear mechanics.
    Scand. Audiol. Suppl. 9, 35-47.
Kemp, D.T. (1981). Physiologically active cochlear micromechanics - one
    source of tinnitus. In: *Tinnitus*, edited by D. Evered and G. Lawrenson
    (Pitman, London), pp. 54-81.
Khanna, S.M. and Leonard, D.G.B. (1981). Basilar membrane tuning in the cat
    cochlea. *Science* 215, 305-306.
Kim, D.O., Neely, S.T., Molnar, C.E., and Matthews, J.W. (1980a). An active
    cochlear model with negative damping in the partition. In: *Psychophysic-
    al, Physiological & Behavioural Studies in Hearing*, edited by G. van den
    Brink & F.A. Bilsen (Delft University Press), pp. 7-14.
Kim, D.O., Molnar, C.E., and Matthews, J.W. (1980b). Cochlear mechanics: non-
    linear behavior in two tone responses as reflected in cochlear-nerve-
    fiber responses and in ear canal sound pressure. *J. Acoust. Soc. Am.*
    67, 1704-1721.
Neely, S.T. (1981). *Fourth-Order Partition Dynamics for a Two-Dimensional
    Model of the Cochlea* Ph.D. Thesis. Univ. of Washington, Saint Louis.
Sellick, P.M., Patuzzi, R. and Johnstone, B.M. (1982). Measurement of basil-
    ar membrane motion in guinea pig using the Mössbauer technique.
    J. Acoust. Soc. Am. 72, 131-141.
Wilson, J.P. (1980a). Evidence for a cochlear origin for acoustic re-emiss-
    ions, threshold fine-structure and tinnitus. *Hearing Research* 2, 233-252.
Wilson, J.P. (1980b). Model for cochlear echoes and tinnitus based on an
    observed electrical correlate. *Hearing Research* 2, 527-532.
Wilson, J.P. (1980c). Model of cochlear function and acoustic re-emission.
    In: *Psychophysical, Physiological & Behavioural Studies in Hearing*,
    edited by G. van den Brink & F.A. Bilsen (Delft University Press),
    pp. 72-73.
Wilson, J.P. and Sutton, G.J. (1981). Acoustic correlates of tonal tinnitus.
    In: *Tinnitus*, edited by D. Evered and G. Lawrenson (Pitman, London),
    pp. 82-107.
Wilson, J.P. and Sutton, G.J. (1983). A family with high-tonal objective
    tinnitus - an update. In: *Hearing - Physiological Bases and Psychophys-
    ics*, edited by R. Klinke and R. Hartmann (Springer, Berlin).
Zurek, P.M. and Clark, W.W. (1981). Narrowband acoustic signals emitted by
    chinchilla ears after noise exposure. *J. Acoust. Soc. Am.* 70, 446-450.

CRITICAL BEHAVIOUR OF AUDITORY OSCILLATORS NEAR FEEDBACK PHASE TRANSITIONS

W.L.C. Rutten, H.P. Buisman

ENT department (KNO), University Hospital
10 Rijnsburgerweg, Leiden, The Netherlands

ABSTRACT

Two classes of cochlear emissors were investigated on their amplitude and phase response to external continuous, sinusoidal stimulation. One class is that of narrowly tuned emissors which become active under stimulation, but show no spontaneous activity. The other class is already spontaneously active and remains active under stimulation. Both classes show non-linear input-output behaviour, frequency "entrainment" and "decoupling" of amplitude- and phase behaviour. Only in the non-spontaneous class the input-output amplitude behaviour can be unambigiously described by power-like behaviour with power p = 2/3. An attempt has been made to compare this power behaviour to that in an electrical near-critical feedback oscillator of the Wien-bridge type and to give an explanation of input-output behaviour in terms of classical Landau theory on "phase transitions and critical phenomena".

1. INTRODUCTION

Since the pioneering work on acoustic emissions (AE's) by D.T. Kemp (1978) various aspects of this phenomenon have been investigated, for example emissive behaviour in normal and pathological ears (Kemp 1978, Kemp and Chum 1980a, Rutten 1980a, 1980b, Schmiedt and Adams 1981, Wilson 1980a, Wit and Ritsma 1979, 1980a, Wit et al. 1981), physiological vulnerability of AE (Anderson and Kemp 1979, Kemp 1982, Siegel and Kim 1982, Zurek and Clark 1982), spontaneous versus evoked AE's (Wilson 1980b, Zurek 1981), implications for models involving non-linear and/or active processes (De Boer 1980, Johannesma 1980, Kemp 1980a, Kim et al. 1980, Rutten 1980b, Wilson 1980c, Wit and Ritsma 1980b, Zwicker, 1979). Experiments and modelwork suggest that AE's originate in very narrowly tuned preneural active vibratory sources in the Organ of Corti. The vibratory sources can be either spontaneously active or become active upon stimulation, either *after* transient stimulation or *during* continuous sinusoidal stimulation. Sometimes, non-spontaneously active sources exhibit "transition behaviour", i.e. they change over to spontaneous behaviour after once having been evoked (sometimes referred to as "triggered tinnitus"). One is tempted to consider that the oscillators in the "black box" of the organ of Corti can be modeled in terms of non-linear feedback amplifiers, of which the feedback factor can be modified externally (by stimulation) or internally (by parameter variation). The feedback factor can be sub-critical (stable passive tuned-filter behaviour), supercritical (stable oscillatory behaviour) or "true critical" (fluctuating oscillatory behaviour).

The idea of studying AE's in terms of "oscillators near feedback transitions" is attractive because electrical realisations of non-linear oscillators have been investigated in the past with regard to their critical properties (for example in Horn et al. 1976, Kawakubo et al. 1973). Moreover, even more attractive is that in these studies it appeared that there exist close analogies between the critical behaviour of oscillatory output voltage on the one hand and the developing of long-range order in many physical systems near phase transitions, such as in transitions in a gas-fluid system, in a paramagnetic-ferromagnetic transition, in ferro-electrics, in lasers near threshold etc. One enters the field of "phase transitions and critical phenomena", a very extensively and successfully studied field in especially solid-state physics during the last two decades (for introduction and review see Stanley 1971 and Domb and Green 1976). Theories in this field, from mean-field or Landau theory to advanced renormalization or mode-coupling theories predict universal behaviour of critical phenomena in widely differing physical systems. The next section gives a very short survey of the critical behaviour of electrical oscillators regarded upon in terms of Landau theory.

2. PHASE TRANSITIONS AND CRITICAL PHENOMENA

Landau theory (see Stanley 1971) basically assumes that in a magnetic system the thermodynamic Helmholtz potential can be expanded in a two-variable Taylor series about the critical temperature $T = T_c$ and about the magnetization $M = 0$. It then predicts power-law behaviour near T_c for quantities such as magnetization, susceptibility, critical isotherm and specific heat. For example in a magnetic phase transition the magnetization M varies with temperature near T_c as $M \sim (T_c-T)^{\frac{1}{2}}$, for field $H = 0$, $T < T_c$ and M small. The isothermal zero-field susceptibility $\chi_T \equiv (\frac{\partial M}{\partial H})_T$ varies as $\chi_T \sim (T-T_c)^{-1}$, for $T > T_c$, $M = 0$ and $H = 0$. At $T = T_c$ the magnetization develops as (the critical isotherm) $M \sim H^{1/3}$. It is conventional to denote the powers as β, $-\gamma$ and $1/\delta$ respectively. Thus $\beta = 1/2$, $\gamma = 1$ and $\delta = 3$ in Landau theory.

Kawakubo et al. (1973) and Horn et al. (1976) showed that a close analogy exists between the above described magnetic transition and the fluctuations of the output voltage of a negative resistance oscillator near threshold of oscillation. Their oscillator was a conventional Wien-Bridge oscillator (Fig. 1). As the circuit equation they use the threshold approximation

$$\frac{d^2v}{dt^2} + 9 \omega_o(\alpha-\alpha_c) \frac{dv}{dt} + \omega_o^2 v = f(t) \tag{1}$$

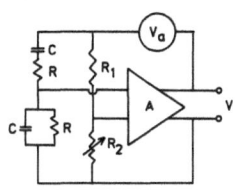

Fig. 1. Wien-bridge electrical oscillator circuit. V_a: external drive. Feedback is controlled by variation of R_2. Non-linearity is introduced by adding non-linear elements to one of the feedback branches.

in which the feedback factor $\alpha = R_2/R_1+R_2$ and $\alpha_{critical} \equiv \alpha_c = 1/3 - 1/A$. As driving force a Johnson noise source $f(t)$, to be thought in V_a, is included. As a non-linear element, Horn et al. add a light bulb to the feedback circuit. Feedback α is then determined by the power dissipated in this non-linear element. The light bulb has a relaxation time τ. This gives

$$\frac{d\alpha}{dt} + \frac{\alpha-\alpha_o}{\tau} = a \ |\Psi|^2 + b|\Psi|^4 + \ldots \tag{2}$$

α_o is the linear part of α, a and b are positive constants, Ψ is the complex amplitude of the oscillatory voltage defined by $V(t) = \mathrm{Re}\ \Psi(t)\ e^{i\omega_o t}$. Stable oscillations then obey the condition

$$\alpha_o - \alpha_c + \tau\, a <|\Psi|^2> + \tau\, b <|\Psi|^4> + \ldots = 0 \tag{3}$$

where $< >$ denote time averages.

When a voltage $V_a = \mathrm{Re}\ E\ e^{i\omega_o t}$ is applied the steady state oscillation condition becomes

$$(\alpha_o - \alpha_c) <|\Psi|> + \tau\, a <|\Psi|^3> + \tau\, b <|\Psi|^5> + \ldots - E/9\omega_o^2 = 0 \tag{4}$$

(in the approximation $f(t) = 0$ one has $<\Psi>^2 = <|\Psi|>^2 = <|\Psi|^2>$).

With the identifications $\alpha_o \rightarrow T$ and $\alpha_c \rightarrow T_c$ and $\Psi \rightarrow M$ and $E \rightarrow H$ Eqs. (3) and (4) are equivalent to those in Landau theory. This implies that the critical behaviour of such an oscillator obeys

$$<\Psi> \sim (\alpha_c - \alpha_o)^{1/2} \qquad\qquad (E = 0,\ \alpha_o \leq \alpha_c) \tag{5}$$

$$\frac{\delta <\Psi>}{\delta E} \sim (\alpha_o - \alpha_c)^{-1} \qquad\qquad (E = 0,\ \alpha_o \geq \alpha_c) \tag{6}$$

$$\text{and } <\Psi> \sim E^{1/3} \qquad\qquad (\alpha_o = \alpha_c) \tag{7}$$

Indeed, Horn et al. did observe experimentally power-law behaviour with powers $\beta = 1/2$, $\gamma = 1$ and $\delta = 3$ in this electrical analogue of a magnetic phase transition (for details see Horn et al. 1976). In order to test Eq. (7) in the AE behaviour of the cochlea one should replace 1) the output voltage Ψ of the electrical oscillator by the output pressure amplitude P_o of an AE (as observed in the outer ear canal) 2) the driving voltage E by the pressure amplitude of a stimulating sinusoidal input P_I (i.e. the input into the ear canal). Unfortunately, the AE equivalent of Eqs. (5) and (6) cannot be tested in humans, as this demands independent variation of the feedback factor α (possibly in an animal model this can be achieved by physiological manipulation). Of course, testing Eq. (7) implies the assumption that the cochlear oscillators are very close to criticallity i.e. $\alpha \approx \alpha_c$.

In the next section we shall investigate to which extent the supposed relation $<P_o> \sim P_I^{1/3}$ is experimentally confirmed.

3. EXPERIMENTAL METHOD AND RESULTS

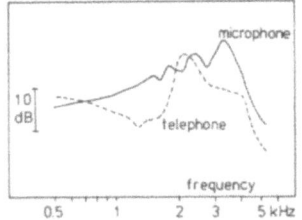

Fig. 2. Drawn line: Amplitude response characteristics of the microphone (Knowles, BT-1751) mounted in the meatal probe device. Dashed line: Amplitude response characteristic of the telephone (Knowles, BK 1606) mounted in the same meatal probe device.

The recording and stimulating equipment consisted of a miniature microphone/telephone assembly, the same as used in previous studies (Rutten 1980b). Their amplitude characteristics are given in Fig. 2. The microphone output voltage was analysed with the aid of a Brookdeal two component lock-in analyser (type 9505 SC) in the two-phase computer mode. The reference input of this analyser was connected to the input of the stimulus controlling amplifier. The stimulus was a continuous sinusoid. Amplitude and phase of the microphone output were analysed as a function of frequency between about 1000 and 4000 Hz. By vectorial analysis of the microphone output phasor the emissor amplitude and phase were separated out of the total signal, which is the vectorial sum of stimulus and emission (see also Kemp 1980b). The frequency of the stimulus was slowly swept back and forth between 1000 Hz and 4000 Hz at various stimulus levels varied by 5 dB steps between -20 dB and +40 dB HL. Also, at the beginning of each experiment, a search was performed for spontaneous emissions by the same procedure as above, but without input to the telephone.

In ten young adults (17 ears) with normal hearing acuity narrowly tuned continuous emissive behaviour was observed. Two types of emissive behaviour could be distinguished. The first type consisted of spontaneously present emissions (fluctuating amplitude , sharply tuned) which became stronger upon external stimulation at their own frequency and also became broader and asymmetrically tuned. The second type consists of emissions which were not spontaneously present but which became active during stimulation at a specific frequency. The main distinction is that their input/output amplitude behaviour (as a group) differs from the first-type group. Four ears showed both types of emissive behaviour. In most ears two or three emissors were observed. In one ear up to eighteen emissors were detected, between 1000 and 3000 Hz with a frequency spacing of about hundred Herz.

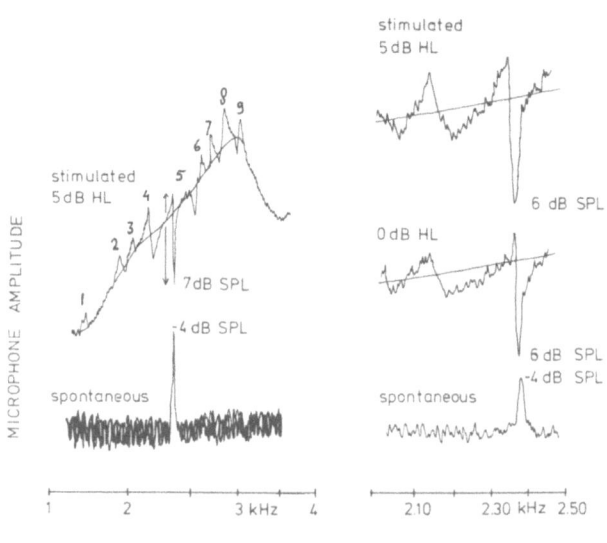

Fig. 3. Left: Microphone output on a linear amplitude scale in subject AS as a function of frequency and stimulus condition. Both spontaneous and stimulated (at 5 dB HL) output are shown. The response is the total response, i.e. the vectorial sum of stimulus and emission. Lower trace: one spontaneous emission with amplitude of 4 dB SPL emerges out of a noise floor of about -16 dB SPL. Upper trace: amplitude of the stimulus is given as the drawn line (see text). The "modulation" (except the large dip) under and above this smooth drawn curve reflects the frequency "entrainment" behaviour of 9 distinct emissions,

which become active under stimulation. *Fig. 3. Right: Same experiments as left, one month later, on an extended frequency scale.*

Figure 3 demonstrates typical results in an ear which has one spontaneous emission at a level of -4 dB SPL at a frequency of 2.38 kHz (lower trace in left part of Fig. 3). The remaining portion of this lower trace shows the noise level in this ear of about -16 dB SPL. The emission is very sharply "tuned", in fact the observed "bandwidth" is that of the two-phase analyser (So, the cochlear emissor may be even more sharply tuned). Upon stimulation with a 5 dB HL sinusoid, (upper trace of left part of Fig. 3), the tuning pattern of this emission changes, it becomes asymmetric, its "peak to valley" level corresponds to 7 dB SPL. In the same trace one observes that nine other emissors become active at the left and right side of the "spontaneous" emission. The smooth drawn curve is the amplitude curve in the absence of emissive behaviour, this curve has been scaled down from the response curve at a stimulation level of 50 dB HL, at which level emissors are saturated and very small compared to the stimulus vector (see also Kemp 1980b). Upon increasing the stimulus level (results not shown) the "amplitude spectrum" retains its shape as at 5 dB HL, but shifted somewhat to lower frequencies. The right part of Fig. 3 shows details of results of remeasurements in the same ear, one month later, presented on an extended frequency scale. A strong similarity between right and left part of Fig. 3 can be observed, indicating the extreme stability of these emissive effects. Whereas Fig. 3 gives the total microphone output, Fig. 4 presents (in an other subject) emissor amplitude and phase, obtained after vectorial subtraction of the input signal from the total signal. This emissor-amplitude plot shows 5 emissors of which one is also spontaneously present at f = 1.61 kHz. At the 15 dB HL stimulus level the

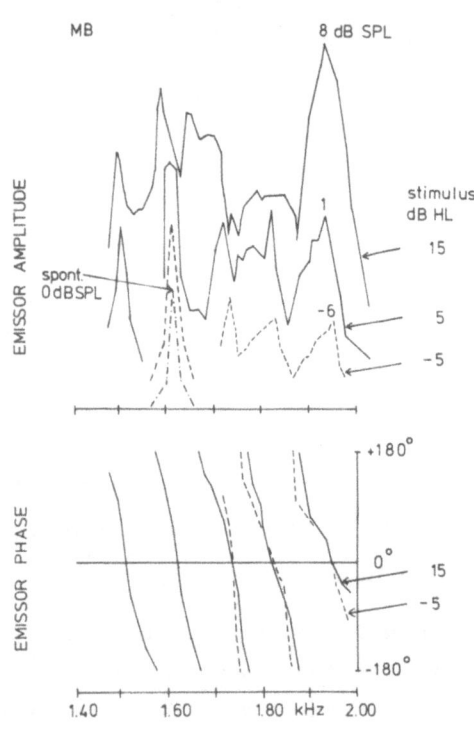

Fig. 4. Lower part: Emissor phase re. stimulus at two stimulus levels, -5 and 15 dB HL. The subsequent downward sloping curves are in fact continuously connected to each other. Phase behaviour remains unchanged under stimulus level variation. Upper part: Emissor amplitude at three stimulus levels (dashed and drawn curves) and spontaneous (dashed-dotted curve). The vertical scale is linear, however, curves are arbitrarily shifted vertically. At three peaks dB SPL is indicated. Note that peaks shift gradually to lower frequencies upon increasing the stimulus level.

amplitude spectrum shows broadened tuning. The frequency of each peak is "entrained" to the left with increasing stimulus level. Corresponding phase spectra (lower part of Fig. 4) do not show this entrainment, i.e. emissor phase behaviour is unchanged under stimulus variation.

Figure 5 summarizes the input-ouput (I/O) amplitude behaviour of emissors (across all ears) which were present under stimulation but which disappeared when the stimulus was removed. So, Fig. 5 presents the I/O characteristics of the non-spontaneous class of oscillators. The curves have been arbitrarily shifted along the horizontal axis in order to get a more clear overall view. For the

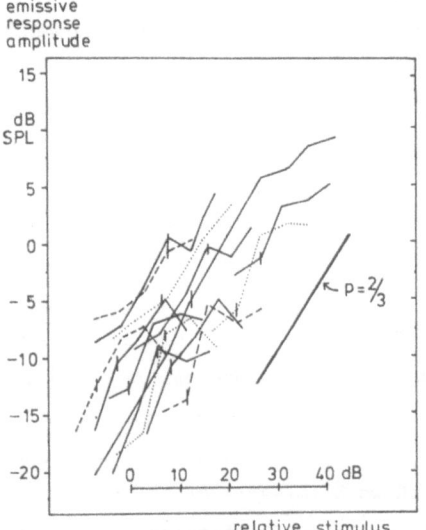

same reason, some of the curves have been dashed or dotted. The short vertical mark on each curve indicates the 5 dB HL stimulus level. The average input-output characteristic can be described by a power-law with power $p = 2/3$. About half of the number of the emissors show saturation behaviour. One of them shows power-like be-

Fig. 5. Non-spontaneously present emissors. Input-output behaviour of emissor response under stimulation at the emissor frequency (at maximum of tuning). Curves have been shifted arbitrarily along the stimulus scale. Vertical short lines at each curve denote 5 dB HL stimulus level. Average I/O power can be described very well by $p = 2/3$.

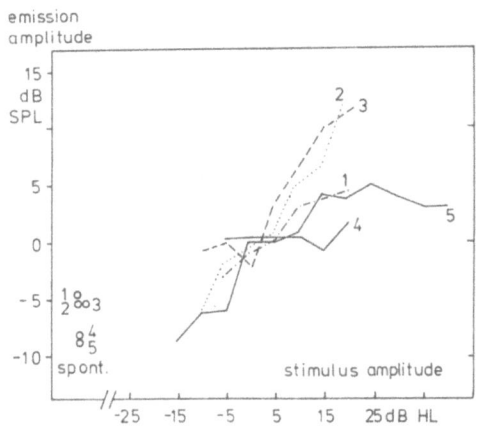

Fig. 6. Same as fig. 5, but for spontaneously present emissors. The spontaneous emissor amplitude is indicated at left-under. Upon stimulation one finds the curves on the right. Horizontal scale in dB HL, curves have not been shifted as in fig. 5. I/O behaviour varies widely from p = 2/3 to gradually sloping (curve 5 for example) and finally to full saturation i.e. p = 0.

haviour over more than 40 dB, which is an exceptionally wide range. For the other class of emissors, i.e. the spontaneous class, input/output results are given in Fig. 6. The open circles at the left show the spontaneous amplitude, while the output under stimulation is given in the various traces on the right. Numbers correspond to circles at the left. Compared to Fig. 5 the input-output behaviour is less homogenous, power 2/3 is observed (curves 2 and 3) as well as strongly saturating behaviour (curves 4 and 5).

4. DISCUSSION

Experiments under continuous sinusoidal stimulation showed clear-cut power-like I/O behaviour with power p = 2/3 for the non-spontaneous emissors (Fig. 5), while the spontaneous class exhibits no such homogeneous power behaviour (such varying-power like behaviour was also found earlier under transient stimulation see Kemp 1978, Rutten 1980a,b, Schloth 1982, Wit et al. 1981). So, in order to compare with predicted power behaviour, we shall confine ourselves to the non-spontaneous class of oscillators (This confinement also allows to neglect Brownian noise in the cochlea as an initiating factor for spontaneous oscillation, see de Vries, 1956).

Our first concern should be to evaluate whether the condition $\alpha \gtrsim \alpha_c$ is experimentally fullfilled. Indeed, the non-averaged oscillation amplitude fluctuated strongly, with a typical time constant of 5 ms. This was observed in sampled portions of the filtered microphone signal of about 100 ms duration. However, a more precise evaluation of fluctuating behaviour should be in terms of correlation functions, to be performed in future experiments.

The non-linearity of the I/O behaviour together with the "decoupling" of phase and amplitude (as in Fig. 4) indicate that the oscillators belong to the class of non-linear oscillators. Also, the asymmetrical tuning under stimulation (Fig. 3) suggests non-linear limit-cycle oscillation (Machlup and Sluchin 1980). The predicted power $p \equiv 1/\delta = 1/3$ contrasts with the observed power $p = 2/3$. Apart from general criticism on Landau theory (Stanley 1971), one cause for this discrepancy might be that the expansion in Eq. (4) is not appropriate. Modification, for example by changing the positive sign in front of τ a $<|\Psi|>^3$ (Eq. 4) to a negative sign yields $\delta \approx 1$ (see Horn et al. 1976, from their fig. 3 the $<|\Psi|>$ versus E behaviour can be deduced for $\alpha \gtrsim \alpha_c$). So, our experimental result $\delta = 3/2$ lies somewhere "in between" these two expansions. Fitting to the "right" expansion seems achievable. However, construction of the right expansion should be on basis of experimental physiological/physical evidence on the real nature of feedback mechanisms in the organ of Corti, otherwise fitting remains speculative.

Resuming, we think to have presented a first step on a new way in the field of auditory physics to analyse behaviour of acoustic emissions.

Acknowledgements

We thank the Heinsius Houbolt fund for financial support.

REFERENCES

Anderson, S.D. and Kemp, D.T. (1979). The evoked cochlear mechanical response in laboratory primates, a preliminary report. *Arch. Otorhinol.* 224, 47–65.

De Boer, E. (1980). Nonlinear interactions and the "Kemp-echo", *Hear. Res.* 2, 519–526.

Domb, C. and Green, M. (1976). *Phase Transitions and Critical Phenomena* (Academic Press, London).

Horn, P.M., Carruthers, T. and Long, M.T. (1976). Threshold instabilities in nonlinear self-excited oscillators. *Physical Review A* 14, 833–839.

Johannesma, P.I.M. (1980). Narrow band filters and active resonators, a comment. In: *Psychophysical, Physiological and Behavioural studies in Hearing*, edited by G. van der Brink and F.A. Bilsen (Delft University Press), pp 62–63.

Kawakubo, T., Kabashima, S. and Ogishima, M. (1973). Critical slowing down near Threshold of Electrical oscillations. *J. of the Physical Soc. of Japan* 34, 1149–1152.

Kemp, D.T. (1978). Stimulated acoustic emissions from within the human auditory system. *J. Acoust. Soc. Am.* 64, 1386–1394.

Kemp, D.T. and Chum, R. (1980a). Properties of the generator of stimulated acoustic emissions. *Hear. Res.* 2, 213–232.

Kemp, D.T. and Chum, R.A. (1980b). Observations on the generator mechanism of stimulus frequency acoustic emissions – two tone suppression. In: *see under Johannesma*, pp 34–41.

Kemp, D.T. (1982). Cochlear echoes: Implications for noise-induced hearing loss. In: *New perspectives on noise-induced hearing loss*, edited by R.P. Hamernik, D. Henderson and R. Salvi (Raven Press, New York).

Kim, D.O., Neely, S.T., Mölnar, C.E. and Matthews, J.W. (1980). An active cochlear model with negative damping in the partition: comparison with Rhode's ante- and post-mortem observations. In: *see under Johannesma*, pp. 7–14.

Machlup, S., and Sluchin, T. (1980). Driven oscillations of a limit-cycle oscillator. *J. Theor. Biol.* 84, 119–134.

Rutten, W.L.C. (1980a). Evoked acoustic emissions from within normal and abnormal human ears: comparison with audiometric and ECoG findings. *Hear. Res.*

$\underline{2}$, 263-271.

Rutten, W.L.C. (1980b). Latencies of stimulated acoustic emissions in normal human ears. In: *see under Johannesma*, pp. 68-75.

Schloth, E. (1980). Amplitudengung der im äuszeren Gehörgang gemessenen akustischen Antworten auf Schallreize. *Acustica* $\underline{44}$, 239-241.

Schmiedt, R.A. and Adams, J.C. (1981). Stimulated acoustic emissions in the ear canal of the gerbil. *Hear. Res.* $\underline{5}$, 295-305.

Siegel, J.H. and Kim, D.O. (1982). Cochlear Biomechanics: Vulnerability to acoustic trauma and other alterations as seen in neural responses and ear-canal sound pressure. In: *see under Kemp* (1982).

Stanley, H. (1971). *Phase transitions and critical phenomena* (Oxford University Press, London).

de Vries, H. (1956). Physical aspects of sense-organs. *Progr. Biophys. Biophys. Chem.* $\underline{6}$, 236-245.

Wilson, J.P. (1980a). Model for cochlear echoes and tinnitus based on an observed electrical correlate. *Hear. Res.* $\underline{2}$, 527-532.

Wilson, J.P. (1980b). Evidence for a cochlear origin for acoustic reemissions, threshold fine-structure and tonal tinnitus. *Hear. Res.* $\underline{2}$, 233-252.

Wilson, J.P. (1980c). The combination tone, $2f_1-f_2$, in psychophysics and ear canal recording. In: *see under Johannesma*, pp. 43-50.

Wit, H.P. and Ritsma, R.J. (1979). Stimulated acoustic emissions from the human ear. *JASA* $\underline{66}$, 911-913.

Wit, H.P. and Ritsma, R.J. (1980a). Evoked acoustical responses from the human ear: some experimental results. *Hear. Res.* $\underline{2}$, 253-261.

Wit, H.P. and Ritsma, R.J. (1980b). On the mechanism of the evoked cochlear mechanical response. In: *see under Johannesma*, pp. 53-62.

Wit, H.P., Langevoort, J.C. and Ritsma, R.J. (1981). Frequency spectra of cochlear acoustic emissions ("Kemp-echoes"). *JASA* $\underline{70}$(2), 437-445.

Zurek, P.M. (1981). Spontaneous narrow band acoustic signals emitted by human ears. *JASA* $\underline{69}$(2), 514-523.

Zurek, P.M. and Clark, W.W. (1982). The behaviour of acoustic distortion products in the ear canals of chinchillas. *JASA* $\underline{72}$(3), 774-780.

Zurek, P.M. and Clark, W.W. (1981). Narrow band acoustic signals emitted by chinchilla ears after noise exposure. *JASA* $\underline{70}$(2), 446-450.

Zwicker, E. (1979). A model describing nonlinearities in hearing by active processes with saturation at 40 dB. *Biol. Cybernetics* $\underline{35}$, 243-250.

TWO ASPECTS OF COCHLEAR ACOUSTIC EMISSIONS: RESPONSE LATENCY AND MINIMUM STIMULUS ENERGY

H.P.Wit, R.J.Ritsma

Institute of Audiology
University Hospital
Groningen, The Netherlands

ABSTRACT

With time window FFT analysis the latency of cochlear acoustic emissions was calculated. If mainly one frequency is present in the emission signal, this latency is frequency dependent: the lower the frequency, the longer the latency. In the second part of this paper the measurement of minimum stimulus energy to evoke a cochlear acoustic emission is described. This energy is so low (1 eV) that processes at molecular level are likely to play an important role.

1. LATENCY OF COCHLEAR ACOUSTIC EMISSIONS EVOKED BY TRANSIENT STIMULI

The latency of different components in click evoked cochlear acoustic emissions (C.A.E.'s) is frequency dependent. Low frequency components have longer latencies than high frequency components (see for instance the review by Anderson 1980). This frequency dependency is in qualitative agreement with a model proposed by Kemp (1980) for C.A.E.'s. In this model a signal, based on the cochlear travelling wave, is fed back to the stapes. Recent measurements of Johnsen en Elberling (1982) in human subjects however do not show such a clear relationship between latency and frequency. Also measurements in monkeys (Wit and Kahmann 1982) throw some doubt upon the straightforward frequency-latency relationship put forward in earlier papers. Therefore we have remeasured latencies of C.A.E.'s using a time window analysis technique.

We selected one ear (subject LE, right ear) that gives C.A.E.'s at different frequencies when stimulated with a click. Frequencies for relatively strong emissions are: 1.09 kHz, 1.47 kHz, 1.61 kHz and 3.06 kHz. (This ear was also described in Wit et al. 1981). Click evoked C.A.E's were recorded from this ear with standard equipment. (Wit and Ritsma 1979). Clicks were presented with a repetition rate of 40/s. Responses were averaged 1024 times with a Datalab DL 4000 signal averager and analysed with a computer. In order to remove components from the averaged signal that are proportional to stimulus level, two subsequent recordings were subtracted after multiplication of the second recording by a factor of 2. This second recording was made with a 6 dB lower stimulus level as the first recording. (Wit and Kahmann 1982). Frequency spectra of the difference signal obtained in this way were calculated using an FFT algorithm. This was done after multiplication of the difference signal by a 3 ms long Hanning ($\cos^2$) time window. By shifting the centre of this time window along the time axis of the difference signal, the frequency contents of different parts of the signal could be studied. The results of such an analysis of the C.A.E.'s from LE's right ear is given in fig.1. This figure does not show a straightforward relation between response frequency and response latency. In the 3 kHz range for instance responses are present along the entire time axis from 5 to 17 ms after stimulus onset.

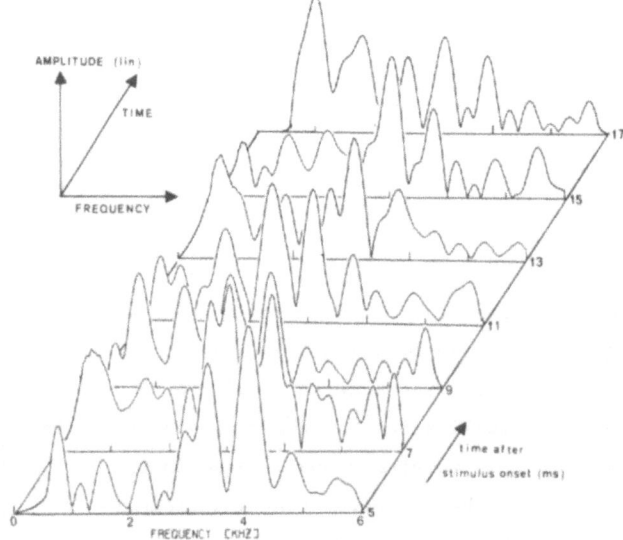

Fig.1. Frequency spectra of parts of C.A.E.signal for different positions of time window.

In earlier measurents we had found (Wit and Ritsma 1980a) that an ear that gives C.A.E.'s at more than one frequency can be selectively stimulated with filtered clicks. By tuning the filter frequency to the frequency of one particular C.A.E. component, this component is more strongly present in the frequency spectrum of the time averaged response signal. Therefore we repeated the procedure described above with filtered clicks as stimuli. The centre frequency of the bandpass filter (24 dB/oct) was set at 1.1, 1.5 and 3.1 kHz in three subsequent series of measurements. The amplitude of the emission component under study is given as a function of time window position in fig.2. Now a clear relation between frequency and latency can be seen for the 3.1 and the 1.5 kHz response components. The relation is less clear for the 1.1 kHz component. Therefore we selected another subject (PK) whose left ear emits predominantly one frequency (1.09 kHz). This ear was stimulated with an unfiltered click and the response signal analyzed. The result of the time window analysis is also given in fig.2. It gives a latency of approx. 11 ms for the 1.1 kHz response. This is at the same position along the time axis where the 1.1 kHz response from subject LE has a maximum. We therefore accept 11 ms as the value for the latency of a 1.1 kHz response.

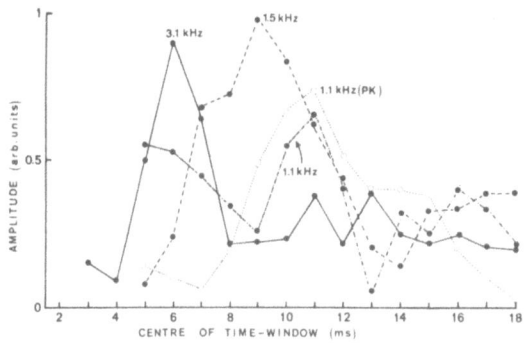

Fig.2. Amplitude of emission components for different time window positions. Black dots represent measurements from the same ear (LE). The open circles are from measurements from another ear (PK).

The latencies that can be derived from fig.2 are plotted in fig.3 as a function of C.A.E. frequency, together with a straight line. This straight line is the best fit to data from earlier measurements (Wit and Ritsma 1980b). The measurements described in this paper give somewhat longer latencies for the 1.5 and the 3.1 kHz response than the values predicted by the straight line. Because this straight line fits latencies of response onset instead of maximum amplitude, this difference is not surprising. A second explanation for the observed latency difference might be that the non-linearity of a response is latency dependent. (Rutten 1980). The longer the latency of a response, the more the relation of its amplitude to stimulus level deviates from linearity. This fact may shift the envelope of a response component towards a longer latency value in the difference signal obtained with the procedure described above. However time window frequency analysis of a single response signal (without subtracting a response signal obtained with different stimulus level) yields exactly the same value for the latency of the 3.1 kHz response (6.1 ms) as that given in fig.3.

Fig.3. Latency of three response components from subject LE's right ear compared to theoretical fit to earlier measurements.

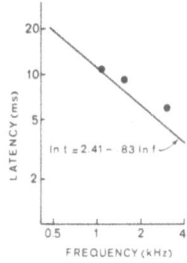

So we conclude that there is a straightforward relation between frequency and latency of isolated C.A.E.'s (predominantly one single frequency). The lower the frequency of the emission, the longer the latency. However for complex emissions (more than one frequency in the response signal) such a simple relation does not seem to be valid.

2. MINIMUM STIMULUS ENERGY

It is well known that stimuli below auditory threshold can still give cochlear acoustic emissions. Wilson (1980a) for instance measured C.A.E.'s with quadrature lock-in amplifiers at stimulus levels 35-50 dB below auditory threshold, with continuous tone stimulation. In order to measure the minimum stimulus energy required to evoke an acoustic response from the cochlea, we selected one subject (DL) who was quiet enough to make measurements of very low response levels possible.

Bandpass filtered (24 dB/oct) 0.1 ms long rectangular pulses were delivered to this subjects right ear with a repetition rate of 40/s. In earlier measurements we had found that this ear has a relatively strong C.A.E. component at 1.29 kHz (Wit et al.1981). Therefore the centre frequency of the stimulus bandpass filter was set at 1.3 kHz. Responses were measured with a sensitive microphone (Wilson 1980b), connected to the ear with a 1 cm long tube. This length of the connecting tube gives the microphone its maximum sensitivity around 1.3 kHz (Wit et al 1981, fig.1). The microphone signal was amplified, high pass filtered (200 Hz) and averaged 2048 times (1024 bins per record; binwidth 20µs). A typical result of a series of 4 measurements is given in fig.4. This figure may give the impression that the response (between dashed lines) can hardly be

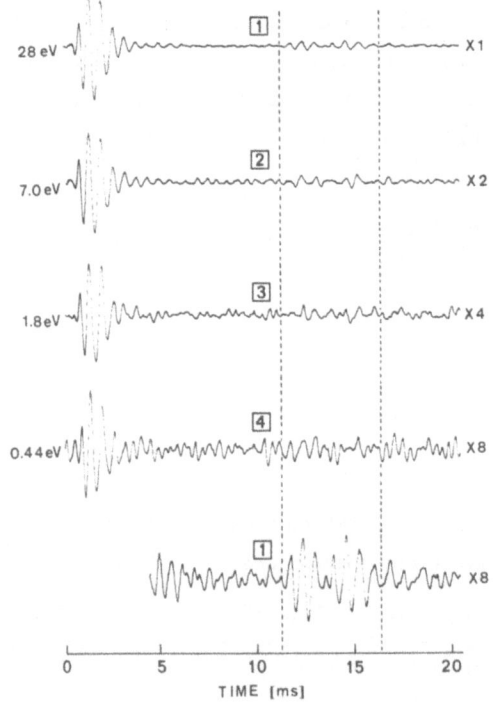

Fig.4. Averaged microphone signal. Stimulus energy decreases with 6 dB steps from top to bottom, while vertical scale amplification increases with 6 dB steps. Response component is present between dashed lines.

separated from background noise. However, stimulus levels are already very low
in fig.4 (the 28 eV stimulus was approx. 10 dB below audible threshold). Stimulus energy was calculated by squaring the microphone signal and integrating it
over the first 4.5 ms. Transformation into electron volts of energy entering
the middle ear was done by assuming that no energy is reflected at the tympanic
membrane with an area of 70 mm^2. Stimulus peak level was calibrated in a
1.3 cm^3 coupler with a ½" B&K condensor microphone. This coupler had the same
volume as the ear under study, when measured with a clinical middle ear impedance meter (at 220 Hz).

The FFT of the signal interval between dashed lines in fig.4 (256 bins) was
calculated and the amplitude of the 1.17-1.37 kHz interval plotted as a function of stimulus level in fig.5. This amplitude was calibrated by calculating
the FFT of a 1.29 kHz tone burst with constant amplitude in the recording
interval of 256 bins and known sound pressure level.

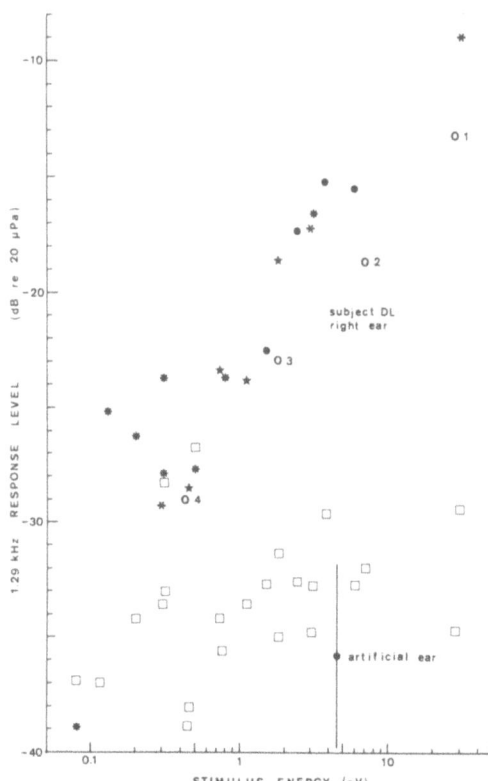

Fig.5. All symbols except open squares: Amplitude of 1.29 kHz acoustic response. Numbered open circles correspond to recordings in fig.4. Open squares represent average amplitude in 0.39-0.98 kHz interval in which no acoustic response is present.

The ear under study permanently emits sound with a frequency of 1.29 kHz (fig.
6). This spontaneously emitted signal is not "locked" to the time scale of the
signal averager if the stimulus is to weak. So in the time averaging process it

will to a large extend disappear (Wit et al. 1981). The spectrum given in fig.
6 was measured with a PAR type 4512 FFT real time spectrum analyser in the
spectrum averaging mode.

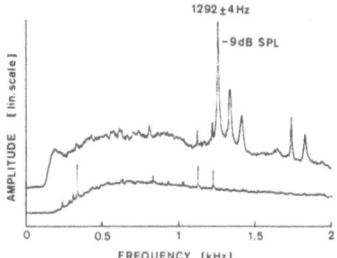

*Fig.6. Top: Spontaneous emission from DL's
right ear.
Bottom: Signal from artificial ear (with a
few 50 Hz harmonics).*

A method complementary to frequency analysis of C.A.E.'s is to make use of the
cross-correlation function for the response signal measured at low stimulus
level and a reference signal measured at a much higher stimulus level (Johnsen
and Elberling 1982). We used a response measured with a 3000 eV stimulus as reference. The maximum cross-correlation coefficient is given as a function of
stimulus level in fig.7.

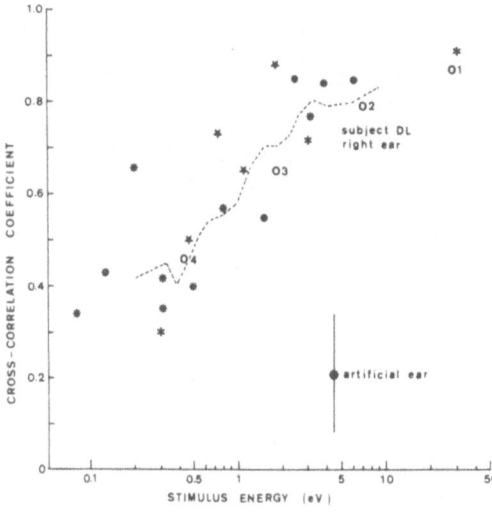

*Fig.7. Maximum cross-correlation
coefficient as a function of stimulus
energy for 1.29 kHz response. Reference signal was a response measured
with high stimulus energy (3000 eV).
The dashed line is the result of a
simple smoothing procedure.*

Figures 5 and 7 show clearly that stimulus energies above approx. 1 eV do influence the permanently present emission at 1.29 kHz in DL's right ear. The
result of this influence is a higher amplitude for the 1.29 kHz component in
the time-averaged response signal than without stimulus. These stimulus energies
are so low that processes at the level of single molecules are likely to be

responsible for the observed phenomena.

This assumption immediately raises the question of how the stimulus energy is concentrated to reach the molecule that absorbs this energy. It is unlikely that "macro"-mechanical processes like vibrations of relatively large parts of the Organ of Corti play an important role. The understanding of absorption of acoustical energy at molecular level in the cochlea also requires a profound study of the influence of Brownian motion upon this process. (De Vries 1952, Harris 1967, Bialek 1980).

Acknowledgements

We thank Jan van Dijk for technical assistance; Lida op 't Ende for typing the manuscript and Meindert Goslinga for improving the figures. This study was supported by the Heinsius Houbolt Fund.

REFERENCES

Anderson, S.D. (1980). Some ECMR properties in relation to other signals from the auditory periphery, *Hearing Res.*2, 273-296.
Bialek, W. (1980). Quantum-mechanical theory of mechanical activity in stereo-cilia and muscle, *J.Acoust.Soc.Am.* 68, S42 (Y).
Harris, G.G. (1968). Brownian motion in the cochlear partition, *J.Acoust.Soc. Am.* 44, 176-186.
Johnsen, N.J. and Elberling, C. (1982). Evoked acoustic emissions from the human ear. II. Normative data in young adults and influence of posture, *Scand.Audiol.*11, 69-77.
Kemp, D.T. (1980). Towards a model for the origin of cochlear echoes, *Hearing Res.*2, 533-548.
Rutten, W.L.C. (1980). Evoked acoustic emissions from within normal and ab-normal human ears: comparison with audiometric and electrocochleographic findings, *Hearing Res.* 2, 263-271.
De Vries, H.L. (1952). Brownian motion and the transmission of energy in the cochlea, *J.Acoust.Soc.Am.* 24, 527-533.
Wilson, J.P. (1980a). Subthreshold mechanical activity within the cochlea, *J.Physiol.*298, 32-33P.
Wilson, J.P. (1980b). Recording of the Kemp echo and tinnitus from the ear canal without averaging, *J.Physiol.*298, 8-9P.
Wit, H.P. and Ritsma, R.J. (1979). Stimulated acoustic emissions from the human ear, *J.Acoust.Soc.Am.* 66, 911-913.
Wit, H.P. and Ritsma, R.J. (1980a). Evoked acoustical responses from the human ear: Some experimental results, *Hearing Res.* 2, 253-261.
Wit,H.P. and Ritsma, R.J. (1980b). On the mechanism of the evoked cochlear mechanical response. In: *Psychophysical, Physiological and Behavioural studies in Hearing*, edited by G. van der Brink and F.A. Bilsen (Delft U.P., Delft), pp. 53-62.
Wit,H.P., Langevoort, J.C. and Ritsma, R.J. (1981). Frequency spectra of cochlear acoustic emissions ("Kemp-echoes"), *J.Acoust.Soc.Am.*70, 437-445.
Wit,H.P. and Kahmann, H.F. (1982). Frequency analysis of stimulated cochlear acoustic emissions in monkey ears, *Hearing Res.* 8, 1-11.

Section IV

Active systems

THE COCHLEAR AMPLIFIER

Stephen T. Neely

*The Boys Town National Institute
for Communication Disorders in Children
Omaha, Nebraska 68131, U.S.A.*

ABSTRACT

The recent observations of sharply-tuned basilar membrane motion and the existence of cochlear acoustic emissions provide evidence that the cochlea is an active mechanical system, capable of generating mechanical vibrations. A model of cochlear mechanics is presented in this paper to support the hypothesis that the primary function of mechanical generators in the cochlea is to amplify displacements of the basilar membrane at sound levels near the threshold of hearing. Some numerical solutions of this model show basilar membrane displacement amplitudes of about 1 angstrom for sound pressures at the eardrum of 0 dB SPL (20 µPa). The rate of energy flow out of the basilar membrane into the cochlear fluid due to the cochlear amplifier is often more than 40 dB greater than the rate of energy flow into the cochlea from the stapes. The action of the cochlear amplifier in this model may be interpreted as a piezoelectric effect which provides a delayed positive feedback to basilar membrane motion.

1. INTRODUCTION

The experimental observations in recent years of cochlear acoustic emissions (Kemp, 1978) and sharply-tuned basilar membrane (BM) displacements (Sellick et al., 1982) have forced us to modify some traditional ideas about cochlear mechanics. The existence of acoustic emissions from the cochlea indicates the presence of mechanical generators within the cochlea. If we consider the cochlea to be an active mechanical system which can utilize available biochemical energy to generate mechanical vibrations, then we can begin to explain the sharply-tuned frequency response observed in the firing rate of auditory nerve fibers directly in terms of BM displacements. A model of cochlear mechanics is presented in this paper to support the hypothesis that the primary biological function of mechanical generators in the cochlea is to amplify BM displacements at sound levels near the threshold of hearing.

2. THE COCHLEAR MODEL

We will consider a linear, two-dimensional, ideal-fluid model of cochlear mechanics (Neely, 1980). The two fluid-filled chambers are each of height H in the y dimension and length L in the x dimension; they are separated by a cochlear partition at y=0, which is open in the apical region $(L-L_h) < x < L$ to represent the helicotrema. Displacements of the cochlear partition in the model are the same as BM displacements. The stapes boundary at x=0 drives the cochlear fluid.

The cochlear amplifier will be implemented by modifying the (usually passive) representation of the mechanics of the cochlear partition. Since we are dealing with a linear model, it is convenient to characterize the partition mechanics by a complex-valued driving-point impedance Z. The driving-point impedance is defined as the ratio of pressure difference across the cochlear partition to BM velocity as a function of position x and radian frequency $\omega = 2\pi f$. The units of Z are $dyn\cdot sec\cdot cm^{-3}$.

The BM impedance Z will be defined as the sum of two parts

$$Z(x,\omega) = Z_b(x,\omega) + Z_a(x,\omega) \tag{1}$$

where Z_b represents the contribution of the mass, damping, and stiffness of the BM

$$Z_b(x,\omega) = i\omega M_1(x) + R_1(x) + K_1(x)/i\omega \tag{2}$$

and Z_a represents the effect of the cochlear amplifier

$$Z_a(x,\omega) = \frac{-(R_3)^2}{i\omega M_2(x) + R_2(x) + K_2(x)/i\omega} \cdot \tag{3}$$

The definition of Z_a was chosen according to the restriction that the cochlear amplifier should add only one degree-of-freedom to the partition mechanics at each position. The physical interpretation for Z_a is that it represents an active biomechanical system located in the vicinity of the outer hair cells, which exerts a force on BM in parallel with the force due to fluid pressure. An alternative definition for Z_a will be presented in section 4.

The choice of numerical values for BM impedance is largely a trial-and-error process relying on physical principles and comparisons between numerical solutions of the model and experimental observations. The following mechanical parameter functions were chosen to represent a cat cochlea using as a guide the cochlear input impedance of Lynch et al. (1982) the cochlear frequency-to-place map of Liberman (1982) and the single nerve fiber response measurements of Allen (1983):

$$K_1(x) = 10^9 \exp(-2.4x) \tag{4}$$

$$R_1(x) = 400 + 10^3 \exp(-1.2x) \tag{5}$$

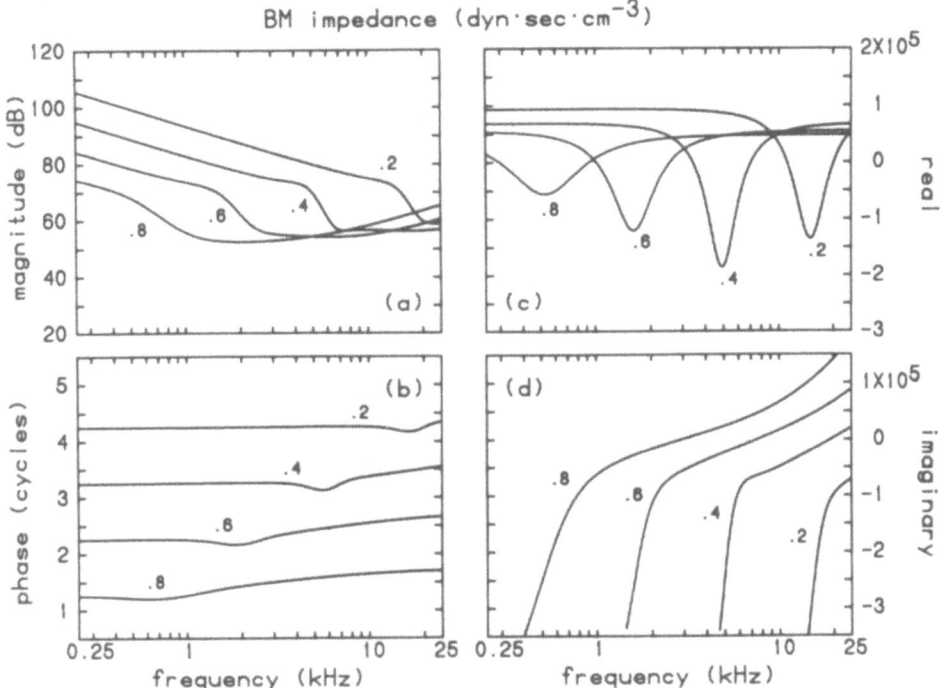

Fig. 1. The driving-point impedance Z of the basilar membrane (BM) is shown as a function of frequency at four places. The numerals next to each curve indicate the position as a fractional distance along BM. Part (a): magnitude of Z in dB re 1 dyn·sec·cm⁻³; part (b): phase of Z with π radian (1/2 cycle) per division; part (c): real part of Z; part (d): imaginary part of Z. The impedance was computed at frequencies which are multiples of 48.8 Hz using the equations for Z given in the text.

$$M_1(x) = 10^{-3} \exp(1.2x) \tag{6}$$

$$K_2(x) = 250 \exp(-2.2x) \tag{7}$$

$$R_2(x) = 10^{-4} \exp(1.1x) + 8\times10^{-4} \exp(-2.2x) \tag{8}$$

$$M_2(x) = 3\times10^{-9} \exp(2.2x) \tag{9}$$

$$R_3(x) = 1.0 \tag{10}$$

The BM impedance is shown in Fig. 1 at four positions: $x/(L-L_h) = 0.2$, 0.4, 0.6, and 0.8, where $(L-L_h) = 2.5$ cm is the length of BM. The effect of Z_a on the BM impedance is clearly seen in Fig. 1(c). At each BM position there is a resonant frequency for Z_a, at which the real part of Z_a, which is always negative, reaches its largest (most negative) value. For the parameters chosen, the real part of Z_a becomes sufficiently negative to cause the real part of the driving-point impedance Z to become negative as well for a certain range of frequencies for each BM position. Whenever Z has a negative real part, the BM is a source of mechanical energy; this energy flows out at BM and into the cochlear fluid. The energy flow out of the BM does not cause the cochlear mechanical system to become unstable, if there is sufficient damping in other regions of BM to absorb the generated energy.

3. NUMERICAL SOLUTIONS FOR BM DISPLACEMENT

Numerical solutions for the cochlear model were obtained by a time-domain, finite-difference method (Neely, 1981). The height H = 0.1 cm and length L = 2.55 cm were represented with 6 and 409 points respectively. The stimulus was defined as a low-pass click voltage, followed by simple models for earphone transducer and middle-ear (Matthews, 1980). The cochlear state was computed at 1 microsecond intervals for 20480 time-steps. The computed eardrum pressure and BM displacement at 4 positions were saved every 10 time-steps. The frequency response of BM displacement re eardrum pressure (shown in Fig. 2) was computed as the ratio of the discrete Fourier transforms of the respective 20 msec time responses. The frequency interval between data points is 48.8 Hz.

The amplitude of BM displacement in Fig. 2(a) is normalized to show the peak amplitude in dB re 1 angstrom (10^{-10}m) corresponding to 0 dB (re 20 μPa rms) sound pressure level at the eardrum. The shape and characteristic frequency of the four response curves are similar to that of typical single nerve fiber responses in cat. The amplitude of BM displacement relative to eardrum pressure in this model at 7 kHz is nearly the same as that measured by Sellick et al. (1982) in guinea pig at 18 kHz. It should be noted that the amplitude ratio shown in Fig. 2(a) would probably be larger if the model were three-dimensional and could represent BM width as being less than the entire width of the cochlea.

The slope and delay functions in Figs. 2(c) and 2(d) were computed by taking differences of adjacent points of the amplitude and phase functions, respectively. The delay function represents the slope of the phase in cycles per kHz. The "stair-step" appearance of the delay curves in Fig. 2(d) is an indication of two straight-line segments in the phase curve plotted on a linear frequency scale. This was also a characteristic feature of Rhode's (1978) observations of the phase of BM displacement in squirrel monkey.

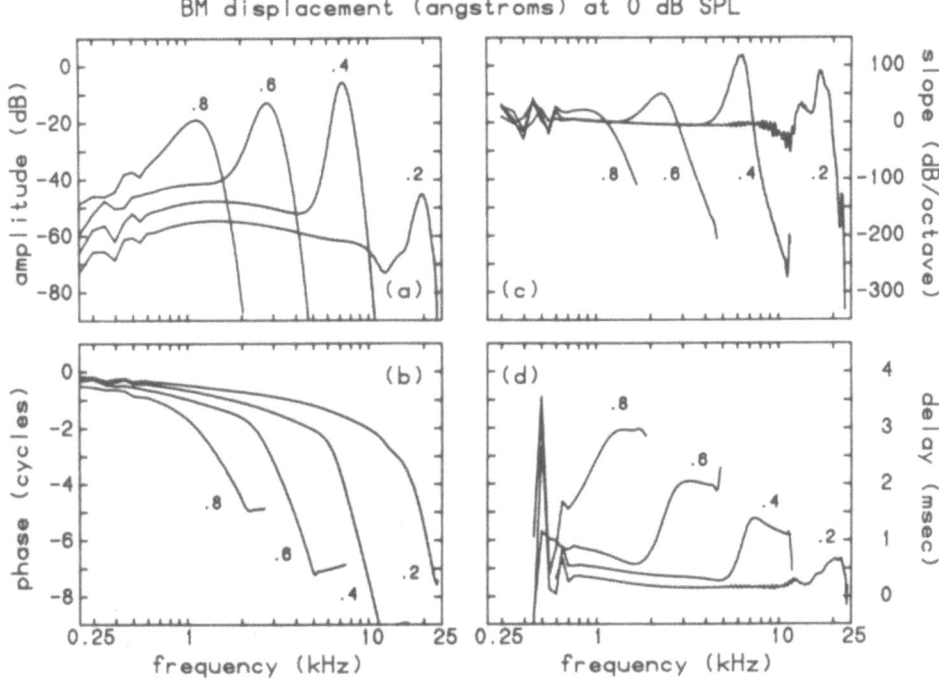

Fig. 2. Displacement of the basilar membrane (BM) re sound pressure at the eardrum. Part (a): peak amplitude of BM displacement in dB re 1 angstrom for eardrum pressure of 20 μPa rms; part (b): phase of BM displacement re eardrum pressure with 1 cycle per division; part (c): slope of the magnitude in dB/octave; part (d): delay (slope of the phase) in msec.

An attractive feature of the cochlear amplifier hypothesis is that it provides a means of explaining the deterioration of sharp tuning which is often observed experimentally in a traumatized cochlea. The loss of the tip of a tuning curve is interpreted as a loss of cochlear amplifier gain. The impedance parameter R_3 can be considered as a sort of gain control on the cochlear amplifier. Solutions of the cochlear model for R_3 = 1.0, 0.9, and 0.0 are shown in Fig. 3. The decreased cochlear amplifier gain clearly has a significant effect on BM displacement near the characteristic frequency. The amplitude at 7 kHz in Fig. 3(a) drops about 20 dB for R_3 = 0.9 and drops about 70 dB for R_3 = 0.0.

4. DISCUSSION

The BM impedance presented in section 2 can be "synthesized" by a mechanical system consisting of 2 masses, 2 springs, 2 positive damping elements, and 1

BM displacement (angstroms) at 0 dB SPL

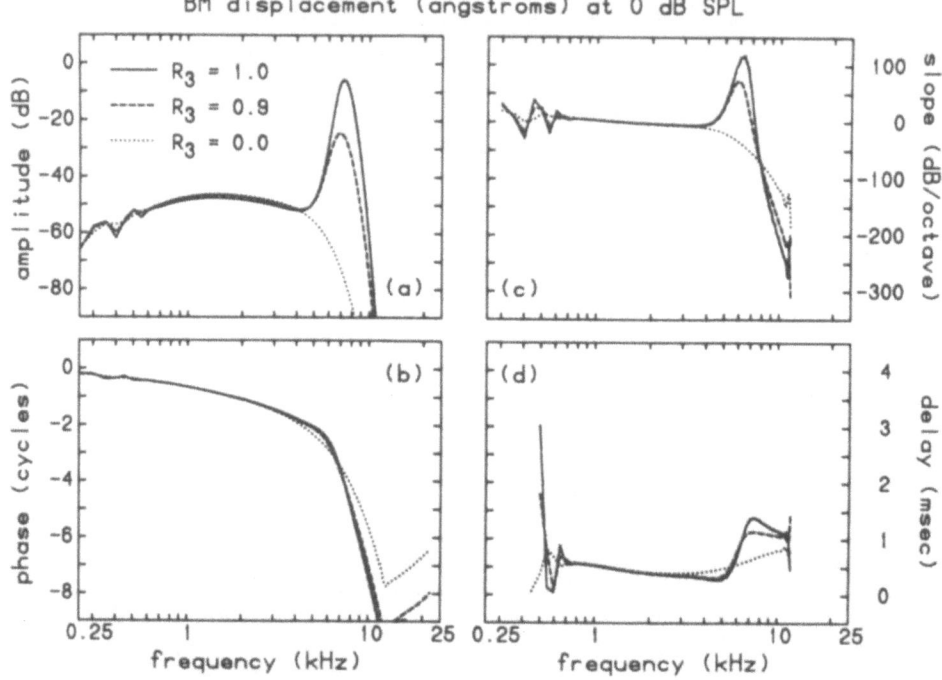

Fig. 3. Effect of changes in cochlear amplifier gain on basilar membrane
(BM) displacement. Parts (a) - (d): same as in Fig. 2. The solid lines
are reproduced from Fig. 2 (position 0.4, R_3 = 1.0). The dashed lines
show BM displacement with R_3 = 0.9 and the dotted lines show BM displacement with R_3 = 0.0.

negative damping element. The use of negative damping in BM impedance was
first presented as a means of explaining the ante- to post-mortem changes in
BM displacement (Kim et al., 1980). A cochlear model with negative damping
elements can produce BM displacements which closely resemble typical neural
responses (Neely and Kim, 1983). Negative damping provides a convenient
means of modeling active mechanical behavior in a linear cochlear model.

A more appealing physical interpretation of the cochlear amplifier is that
it represents a piezoelectric action powered by the cochlear microphonic
(Davis, 1981). This type of electromechanical action can be modeled as a
stiffness component which has a delayed effect. The corresponding definition
for Z_a would be

$$Z_a(x,\omega) = [K_3(x)/i\omega] \exp[-i\omega \tau(x)] \tag{11}$$

where τ is a transduction latency of several microseconds. Preliminary frequency-domain model results using Eq. (11) to implement the cochlear amplifier are very similar to those using Eq. (3). Time-domain implementation, however, is much more difficult for Eq. (11).

Non-linearities in cochlear mechanics are likely to originate in the cochlear amplifier. One plausible way of modeling a nonlinear cochlear amplifier would be to set R_3 (or K_3) to zero if BM displacement exceeds some threshold, thus simulating saturation of the cochlear amplifier.

Evidence for the existence of the cochlear amplifier is continually accumulating (Davis, 1983). Both analytical (de Boer, 1983) and numerical (Neely and Kim, 1983) model results now indicate that a cochlear amplifier is essential for producing basilar membrane displacements with the high sensitivity and sharp tuning observed experimentally.

REFERENCES

Allen, J.B. (1983). Magnitude and phase frequency response to single tones in the auditory nerve, *Journal of the Acoustical Society of America*, (in press).
de Boer, E. (1983). On active and passive cochlear models -- toward a generalized analysis, *Journal of the Acoustical Society of America* (in press).
Davis, H. (1981). The second filter is real, but how does it work? *American Journal of Otolaryngology* 2, 153-158.
Davis, H. (1983). An active process in cochlear mechanics, *Hearing Research* 9, 79-90.
Kim, D.O., Neely, S.T., Molnar, C.E., and Matthews, J.W. (1980). An active cochlear model with negative damping in the cochlear partition: Comparison with Rhode's ante- and post-mortem observations. In: *Psychological, Physiological, and Behavioral Studies in Hearing*, edited by G. van den Brink and F.A. Bilsen (Delft University Press, Delft, the Netherlands), pp. 7-14.
Liberman, M.C. (1982). The cochlear frequency map for the cat: Labeling auditory-nerve fibers of known characteristic frequency, *Journal of the Acoustical Society of America* 72, 1441-1449.
Lynch, T.J., Nedselnitsky, V., and Peake, W.T. (1982). Input impedance of the cochlea in cat, *Journal of the Acoustical Society of America* 72, 108-130.
Matthews, J.W. (1980). *Mechanical Modeling of Nonlinear Phenomena Observed in the Peripheral Auditory System*. Doctoral dissertation (Washington University, St. Louis, Missouri).
Neely, S.T. (1980). Finite difference solution of a two-dimensional mathematical model of the cochlea, *Journal of the Acoustical Society of America* 69, 1386-1393.
Neely, S.T. (1981). *Fourth-order Partition Mechanics for a Two-dimensional Cochlear Model*. Doctoral dissertation, (Washington University, St. Louis, Missouri).
Neely, S.T., and Kim, D.O. (1983). An active cochlear model shows sharp tuning and high sensitivity, *Hearing Research* (in press).

Rhode, W.S. (1978). Some observations on cochlear mechanics, *Journal of the Acoustical Society of America* 64, 158–176.

Sellick, P.M., Pattuzzi, R., and Johnstone, B.M. (1982). Measurement of basilar membrane motion in the guinea pig using the Mössbauer technique, *Journal of the Acoustical Society of America* 72, 131–141.

ELECTROMECHANICAL PROCESSES IN THE COCHLEA

D.C. Mountain, A.E. Hubbard, T.A. McMullen

Boston University
Boston, MA   02215
U.S.A.

ABSTRACT

We propose a feedback model for cochlear mechanics in which the hair cell cilia exert an active restoring force in response to displacement. The restoring force is described by first order kinetics and leads to an increase in sensitivity and frequency selectivity of the basilar membrane response. Several different experimental results from our laboratory suggest that hair cell membrane potential plays an important role in this feedback process. They include the measurement of acoustic emissions in response to passing sinusoidal electric current through the cochlea and the modulation of the sound pressure level (SPL) of single tones by electrical current. The latter effect saturates at levels of the acoustic stimulus similar to the level at which the cochlear microphonic saturates. The experimental data suggest that saturation of the transduction process at high SPL could account for the saturating nonlinearity observed by others in basilar membrane displacement. These electromechanical effects may also explain the generation of acoustic harmonics and distortion products as well as the positive cochlear summating potential.

1. INTRODUCTION

We have been exploring an extension of a hypothesis of cochlear function originally proposed by Gold (1948) and investigated in more detail by Neely and Kim (1982). The important assumption is that the outer hair cells exert an active force on the basilar membrane in response to movement of their cilia. Our model differs from that of Gold in that the active force is proportional to displacement rather than velocity. The model is a negative feedback system since at low frequencies the action of the hair cells is to oppose the movement of the basilar membrane. Figure 1 is a block diagram of the model in which the output of the hair cell displacement transducer drives the force generating process. The coupling between transduction and force generation may be mediated either by receptor potential or receptor current, but we will assume for purposes of this paper that the receptor potential is the important variable.

120

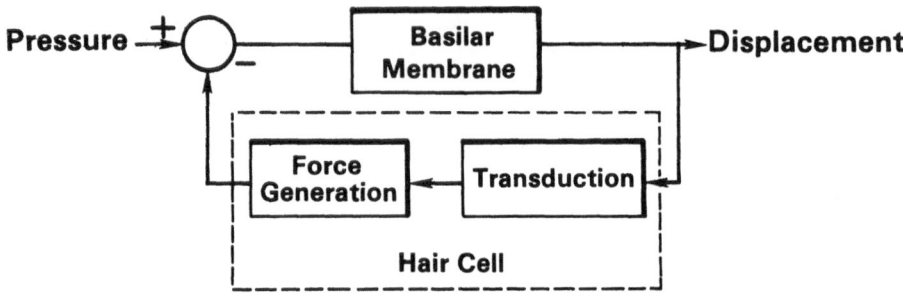

Fig. 1. Block diagram of the cochlear model. The feedback loop represents influence of the hair cells on the basilar membrane.

2. MODEL OF OUTER HAIR CELL CILIA

We have assumed for simplicity that the hair cell cilia can be modeled as a voltage-dependent spring. The spring constant K is a linear function of the hair cell membrane potential $v(t)$:

$$K = K_o(V_o + v(t)) \qquad (1)$$

where K_o and V_o are constants. We further assume that the spring is stretched when the system is at rest so that the relation between the force $f(t)$ and the displacement $x(t)$ with respect to the rest position is given by:

$$f(t) = K_o(V_o + v(t))(X_o + x(t)) \qquad (2)$$

where X_o is the length at rest. Equation (2) can be expanded to give:

$$f(t) = K_o V_o X_o + K_o V_o x(t) + K_o X_o v(t) + K_o v(t) x(t) \qquad (3)$$

We assume that $v(t)$ is a linear function of $x(t)$ for small displacements.

However, v(t) will be low-pass filtered by the hair cell membrane time constant. For larger displacements v(t) is assumed to exhibit a saturating nonlinearity due to the saturation of the transduction process. If we assume small displacements and neglect the fourth term in Eq. (3) then we have a linear relationship between f(t) and x(t) since v(t) is a linear function of x(t). The transfer function relating force to displacement for the hair cell is then:

$$\frac{F(s)}{X(s)} = \frac{K_o K_1 X(s)}{T_m s + 1} \tag{4}$$

where K_1 is the transducer small signal gain and T_m is the hair cell membrane time constant.

3. RESULTS AND DISCUSSION

Figure 2 shows the simulated influence of the hair cells on the transfer function relating basilar membrane displacement to the pressure difference at one point along the membrane. The tectorial membrane in this model is assumed to have an over-damped viscoelastic attachment to the bone. As the gain of the hair cell feedback system is increased there is a dramatic increase in the basilar membrane response near the resonant frequency. In addition there is a modest reduction in the response for frequencies somewhat below the resonant frequency. This reduction in response has been observed by Khanna and Leonard (1982).

The hair cell receptor potential is known to saturate as sound level is increased. Consequently we assume that the feedback gain provided by the hair cells will decrease at higher SPLs causing a decrease in the basilar membrane sensitivity at high SPLs. This behavior has been observed by Rhode (1971).

122

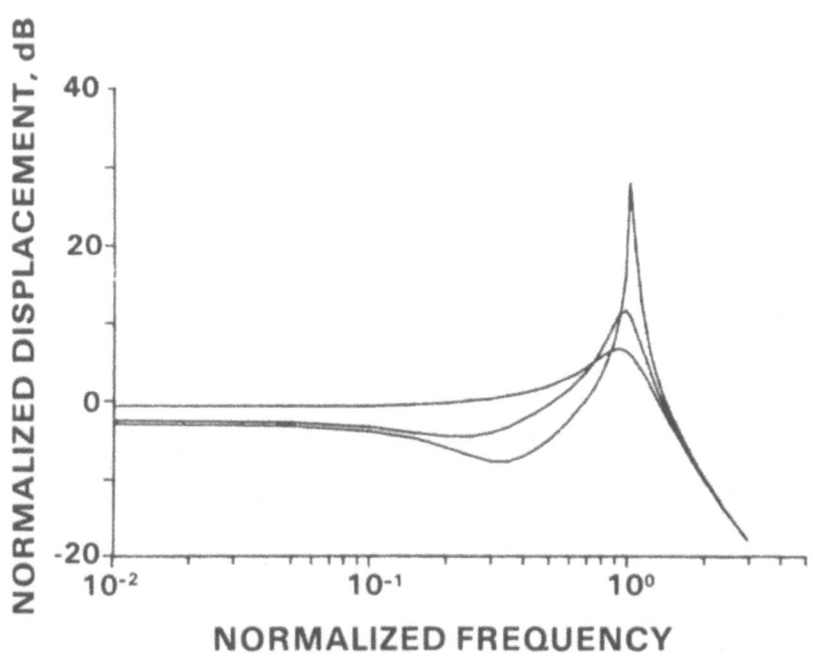

Fig. 2. Magnitude of the basilar membrane transfer function as a function of frequency. The increasing amplitude of the resonant peak results from increasing the feedback gain.

The nonlinearity of the transduction process as well as the fourth term of Eq. (3) leads to the production of harmonic and intermodulation distortion. This nonlinearity also suggests a possibility for the origin of the positive summating potential (SP) provided the voltage-dependent stiffness makes the cilia stiffer when displaced in the depolarizing direction than when displaced in the hyperpolarizing direction. If we apply a constant sinusoidal pressure to the hair cell described by Eq. (3) and couple it to an electric circuit model of the cochlea, then both positive and negative summating potentials can be produced (McMullen, 1983). The positive SP is the result of the voltage-dependent stiffness and the negative SP is the result of the asymmetrical saturation of the transducer. The positive SP dominates the extracellular response at low frequencies and the negative SP dominates at high frequencies. The simulated intracellular SP is negative at low frequencies and positive at high frequencies. This result has been shown by Dallos et al. (1982).

Equation (3) suggests that if we could independently vary V(t), then the hair cells should exert a proportional force on the basilar membrane and in turn on the cochlear fluids. The resulting pressure changes then would propogate back to the stapes and out through the middle ear to the tympanic membrane.

To study this possibility we have injected sinusoidal current into scala media in the gerbil in an attempt to vary v(t). We simultaneously measured the sound pressure changes in the external meatus near the tympanic membrane. We found components in the sound spectrum at the electrical frequency and its harmonics (Mountain and Hubbard, 1983). In addition, if an external tone was applied to the ear, then modulation products (sidebands) were observed at the sum and difference of the acoustic and electrical frequencies (Hubbard and Mountain, 1982). Figure 3 shows the spectrum from such an experiment with the external tone present. Both the component due to the current injection alone and the sidebands disappear after the death of the animal and both can be eliminated by acoustic trauma.

The component at the electrical frequency can be enhanced by the presence of an external tone (see Fig. 4a). Both the enhancement of the electrical component and the magnitude of the sidebands (Fig. 4b) are increased as the level of the external tone is increased. They tend to show saturation at high sound levels. This saturation occurs at levels of the external tone for which the cochlear microphonic saturates (see Fig. 4c).

Our results suggest an important role for the transduction process in the hypothesized feedback loop. If this is indeed the case then any experimental manipulation which alters the sensitivity of the transduction process will

124

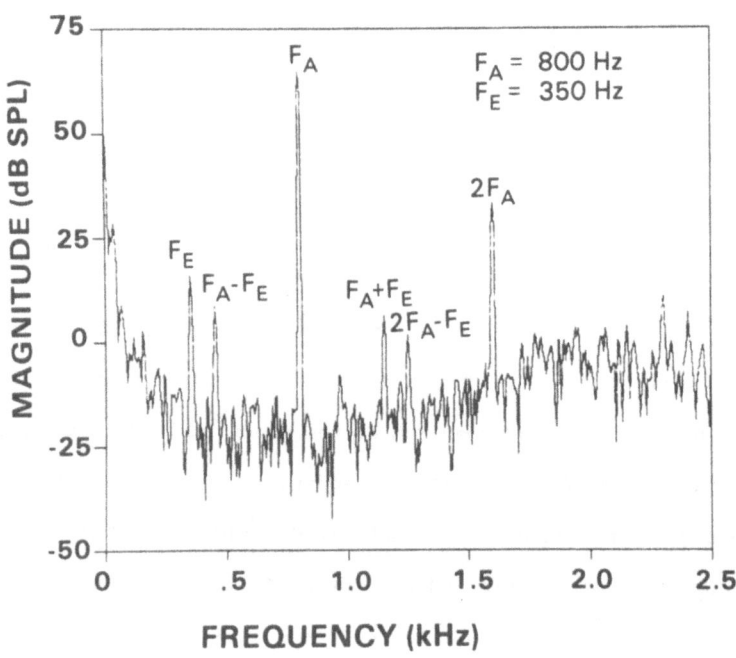

Fig. 3. The acoustic spectrum measured at the tympanic membrane in response to a 350 Hz (F_E) 10 uA p. current and an external tone at 800 Hz (F_A).

have a major impact on cochlear mechanics. For instance a decrease in the endocochlear potential (EP) will reduce the driving force on ionic flow through the transduction channels and in turn decrease the feedback gain. Hypoxia causes a decrease in EP and also causes a dramatic change in auditory nerve fiber sensitivity and frequency selectivity (Evans, 1976). Reduction of the EP by dc current injection alters cochlear nonlinearity (Mountain, 1980) and alters the tuning and sensitivity of inner hair cells (Nuttall, 1983) and of auditory nerve fibers (Hubbard et al., 1983 ; Teas et al., 1970).

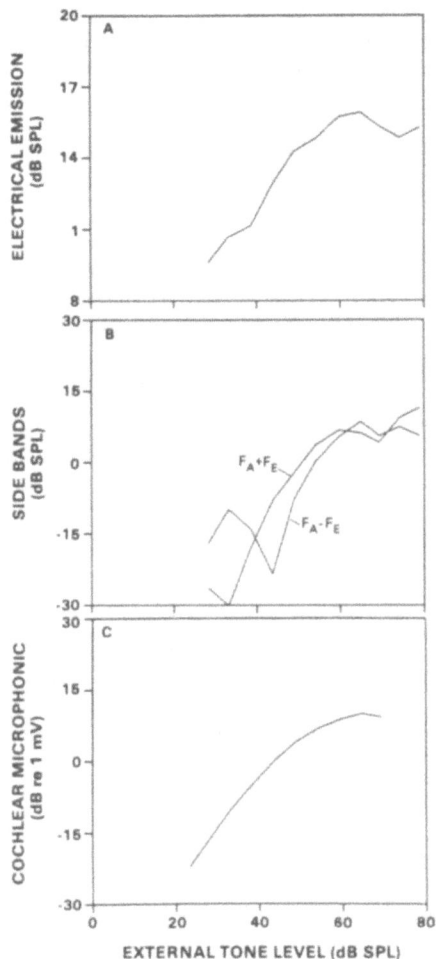

Fig. 4. The response to an external tone at 800 Hz of; (a) the electrical emission, (b) the sideband emissions, (c) the cochlear microphonic. In (a) and (b) the current level was 10 uA p. at 350 Hz.

REFERENCES

Dallos, P., Santos-Sacchi, J., and Flock, Å. (1982). Intracellular recordings from cochlear outer cells, *Science* 218, 582-584.

Evans, E.F. (1976). Temporary sensorineural hearing losses and 8th nerve changes. In: *Effects of Noise on Hearing,* edited by D. Henderson, R.P. Hamernik, D.S. Dosanjh and J.H. Mills (Raven Press, New York), 199-221.

Gold, T. (1948). Hearing. II. The physical basis of the action of the cochlea, *Proc. Roy. Soc. B.* 210, 71-72.

Hubbard, A.E., and Mountain, D.C. (1982). Injection of AC current into scala media alters the sound pressure at the tympanic membrane: Variations with acoustic stimulus parameters, *J. Acoust. Soc. Am.* 71, S100.

Hubbard, A.E., Voigt, H.F., and Mountain, D.C. (1983). Injection of direct current into scala media alters auditory-nerve response properties, *Abstracts of the Sixth Midwinter Research Meeting, Association for Research in Otolaryngology,* 103-104.

Khanna, S.M., and Leonard, D.G.B. (1982). Basilar membrane tuning in the cat cochlea, *Science* 215, 305-306.

McMullen, T. (1983). Model of summating potential production by voltage-dependent cilia stiffness, *Abstracts of the Sixth Midwinter Research Meeting, Association for Research in Otolaryngology,* 68-69.

Mountain, D.C. (1980). Changes in endolymphatic potential and crossed olivo-cochlear bundle stimulation alter cochlear mechanics, *Science* 210, 71-72.

Mountain, D.C., and Hubbard, A.E. (1983). Injection of alternating current into scala media produces an ear canal emission of cochlear origin, *Abstracts of the Sixth Midwinter Research Meeting, Association for Research in Otolaryngology,* 103.

Neely, S.T., and Kim, D.O. (1982). An active model for sharp tuning and high sensitivity in cochlear mechanisms, *J. Acoust. Soc. Am.* 71, S16.

Nuttall, A.L. (1983). Inner hair cell dc receptor potential changes from direct currents introduced into the quinea pig cochlea, *Abstracts of the Sixth Midwinter Research Meeting, Association for Research in Otolaryngology,* 104-105.

Rhode, W.S. (1971). Observations of the vibration of the basilar membrane in squirrel monkeys using the Mossbauer technique, *J. Acoust. Soc. Am.* 49, 1218-1231.

Strelioff, D., and Flock, Å. (1982). Mechanical properties of receptor cells in the guinea pig cochlea, *Soc. Neurosci. Abstr.* 8, 40.

Teas, D.C., Konishi, T., and Wernick, J.S. (1970) Effects of electrical current applied to the cochlear partition on discharges in individual auditory-nerve fibers, II. Interaction of electrical polarization and acoustic stimulation, *J. Acoust. Soc. Am.* 50, 587-601.

A NON-LINEAR FEEDBACK MODEL FOR OUTER-HAIR-CELL STEREOCILIA AND ITS IMPLICA-
TIONS FOR THE RESPONSE OF THE AUDITORY PERIPHERY

S. Koshigoe, A. Tubis

Department of Physics, Purdue University, U.S.A.

ABSTRACT

*The direct incorporation of negative resistance and nonlinear damping in a me-
chanical model of the basilar membrane (BM) may be used to correlate BM response data as well as the properties of stimulated and spontaneous emissions
in the external auditory meatus (S. Koshigoe and A. Tubis, 1982). We have
tried to remedy the ad hoc nature of this formal "black box" approach by seek-
ing a microscopic basis for the assumed nonlinear and active BM response.
Such a basis is suggested by evidence that mechanical cochlear function is
strongly influenced by efferent innervation of outer hair cells (OHC) (D.C.
Mountain, 1980; J.H. Siegel and D.O. Kim, 1982). We consider the chain: OHC
stereocilia deflection → nonlinear change in receptor potential (A.J. Hudsneth
and D.P. Corey, 1977)→induced feedback force on the BM as the primary origin
of nonlinear active cochlear response. Our proposed model is qualitatively
compatible with data on cochlear emissions, nonlinear cochlear response, com-
bination-tone psychophysics, and changes in cochlear response induced by COCB
stimulation or variation of the endocochlear potential.*

1. INTRODUCTION

Over thirty years ago, Gold (1948) suggested the possible existence of a coch-

lear biomechanical feedback loop in order to account for the observed frequen-

cy resolution in hearing. He also speculated on the possibility of self-sus-

tained oscillations in these loops which might result in acoustic emissions

of cochlear origin in the external auditory meatus. Recent experimental find-

ings which give some support to Gold's speculations include: 1) sharp neural-

like basilar membrane (BM) tuning at low stimulus values by Khanna and Leo-

nard [1982]; and Sellick et al. [1982]; 2) stimulated ear canal emissions,

first detected by Kemp (1978); 3) spontaneous ear canal emissions found by

Zurek (1981), Wilson and Sutton (1981), and others; 4) variation of nonlinear

cochlear mechanical response via COCB stimulation or ac current injection in

scala media (Mountain, 1980; Mountain and Hubbard, 1982; Siegel and Kim,1982);

and 5) the discovery of the contractile proteins actin and myosin in hair

cell structures (Flock and Cheung, 1977; Tilney, DeRosier, and Mulroy, 1980).

The latter raises the intriguing possibility of a cochlear feedback loop

based on hair-cell stereocilia deflection → nonlinear change in hair-cell po-

tential (Hudspeth and Corey, 1977) → change in mechanical properties of

stereocilia [i.e. a nonlinear feedback force which may have the requisite

properties to account for 1)-4)]. A similar circumstance has been suggested

by Weiss (1982), and by Koshigoe, Kwok, and Tubis (1982).

2. AD HOC MODEL

Our initial attempt (Koshigoe and Tubis, 1982) to formulate a cochlear model to account for some of the properties of ear canal emissions, is based on a one-dimensional second-order (BM) model whose effective resistance R is a function of the BM velocity, $\dot{\xi}_{BM}$, as indicated in Fig. 1(a). The one-dimensionality of the cochlear model allows for the exploration of various parameter ranges with modest computational costs. The model is a merging of previous ones which incorporate nonlinear resistance (Hubbard and Geisler, 1972; Kim, Molnar, and Pfeiffer, 1973; Hall, 1974) and negative damping (Neely, 1981; Neely and Kim, 1983; Koshigoe and Tubis, 1983). In Section 3, a specific cochlear feedback mechanism is proposed as a basis for this type of model. R may become locally very negative as a consequence of a cochlear pathology. This results in retrograde BM waves and related cochlear emissions in the ear canal. The nonlinearity of R serves to limit BM oscillations which may become unstable in linear models (Neely, 1981; Koshigoe and Tubis, 1983). It also may be used to account for external-tone-induced distortion products, frequency shifts, and suppression effects in the BM motion and in the ear canal pressure. Our computational scheme is indicated in Fig. 1(b). The dynamical response of the system is calculated by using the finite difference method for spatial derivatives and applying the modified predictor-corrector method for time integration. In our investigation of spontaneous emissions and external-tone suppression effects, we digital filter the ear canal signal so as to separate the external-tone frequency component from the spontaneous emission component.

As expected, we find stable BM limit cycles which result in spontaneous emissions. Unfortunately, the triggering of these emissions is hard to avoid in this model as long as we adjust the model parameters so as to obtain realistic Kemp echoes. A quenching mechanism is required. We show in Fig. 2, the typical results for the suppression of a 2 KHz, 10 dB SPL spontaneous emission by external suppressor tones of 1 KHz, 3 KHz, and 4 KHz. In agreement with experimental results, we find that the effect of the suppressor is to both reduce and frequency shift the spontaneous emission. For suppressor frequencies less than the spontaneous frequency, the suppression onset occurs at an SPL less than that for a suppressor with frequency greater than the spontaneous frequency. Also the suppression curve slopes downward more sharply for the suppressor frequency less than the spontaneous frequency. The asymmetry of the suppression curves is of course due to the asymmetry of the BM activity pattern for the different suppressor tones. These results are in qualitative agreement with experimental findings (Zurek, 1981; Wilson and Sutton, 1981).

Frequency shifts of the spontaneous emissions were obtained when sufficiently high levels of the suppressor signals were used. Although it was not possible to delineate the precise amounts of shifting because of the resolution of our FFT, the shifts were in qualitative agreement with the results of Wilson and Sutton (1981).

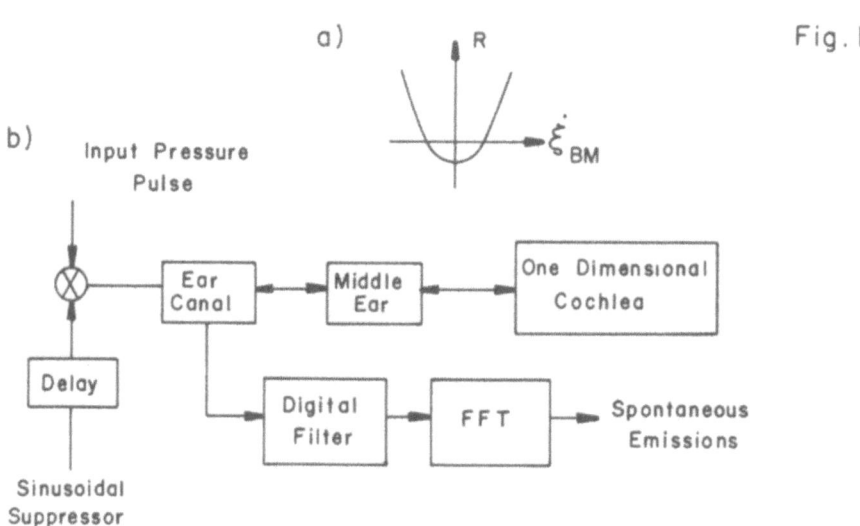

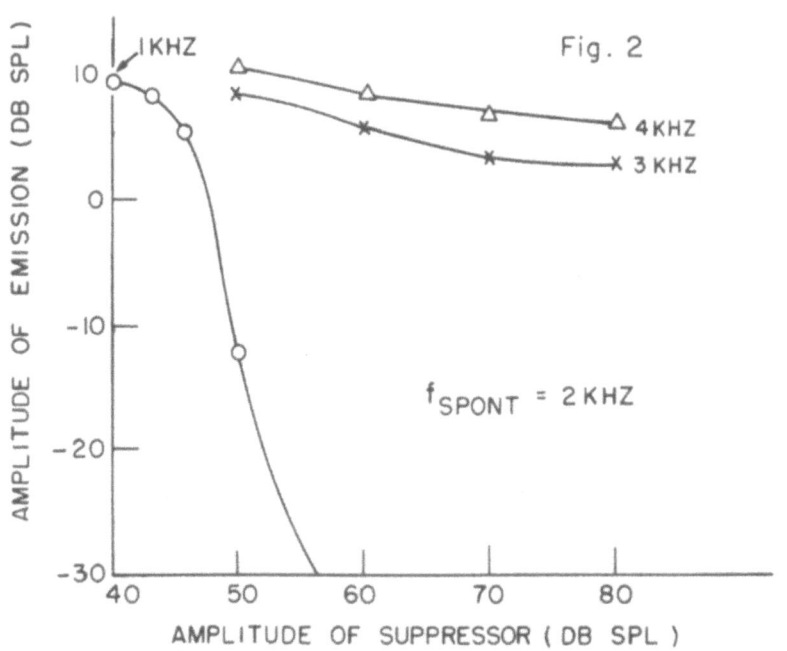

Fig. 1. *Schematics of the ad hoc model for the auditory periphery.*
Part (a): BM resistance as a function of the BM velocity;
part (b) computational set-up. The middle ear model of
Neely (1981) is used.

Fig. 2. *Calculated suppression of a 2 KHz 10 dB SPL spontaneous*
emission by external suppressor tones of 1 KHz, 3 KHz, and
4 KHz. The spontaneous emission level is taken to be that
of ear-canal signal after being passed through a bandpass
filter of center frequency, 2 KHz, and bandwidth, 500 Hz.
The resistance function is $R(d, \dot{\xi}_{BM}) = 250[A(d) + 10^3|\dot{\xi}_{BM}|^2 + 10^6|\dot{\xi}_{BM}|^2]$ *in c.g.s. units, where* d = *distance from*
stapes (cm); A(d) = -1, 1.65 < d < 1.85; A = 0, *otherwise;*
length of cochlea = 2.25 cm.

3. FEEDBACK MODEL

We outline here a possible feedback mechanism (Weiss, 1982; Koshigoe, Kwok,
and Tubis, 1982) which may give some justification for and insight concerning
the validity of the ad hoc type model. In order to make the basic ingredients
of the model clear, we consider for simplicity an outer-hair-cell stereocilium
(OHCS) which is tightly attached at its apex to the tectorial membrane (TM)
(Lim, 1980). Its basal coupling to the rectangular lamina is assumed to give
rise to a force on the BM equal to $-K_o g \xi_{BM} + \Delta f_s$. K_o is the intrinsic stiffness Hooke constant of the coupling, g is the radial shear-lever gain (Rhode
and Geisler, 1967), and Δf_s is the feedback force which will be functionally
related to the time dependence of the OHC membrane potential, $V_H(t)$. The connection between $V_H(t)$ and $x (\equiv g\xi_{BM})$, the OHCS displacement, is assumed to
follow from the hair cell model of Fig. 3(a,b) (Davis, 1965; Widerhold, 1967;
Klinke and Galley, 1974; Weiss, Mulroy, and Altmann, 1974). The x-dependence
of r in Fig. 3(b) is suggested by the results of Hudspeth and Corey (1977).
V_H may be written as $V_H^o + \Delta V_H(t)$, where V_H^o would be the steady state value of
V_H if x(t) = 0 (Weiss, Mulroy, and Altmann, 1974).

Our proposed feedback mechanism is motivated by the observed variation in the
mechanical cochlear response due to either COCB stimulation or induced changes
in the endocochlear potential (EP). (See INTRODUCTION for references.) Both
types of stimulation produce changes in V_H. (COCB stimulation probably decreases R_M by increasing the membrane permeability to ions such as K^+ and
Cl^-.) Thus a strong linkage between V_H and the feedback force, Δf_s, is suggested.

At present, the connection (if any) between stereocilia stiffness and $V_H(t)$ is
unknown. However, Tilney, Egelman, DeRossier, and Saunders (1983),in investigations of the bending of stereocilia in bird cochlea, find sliding of adjacent actin filaments past one another. Changes in V_H may possibly affect the

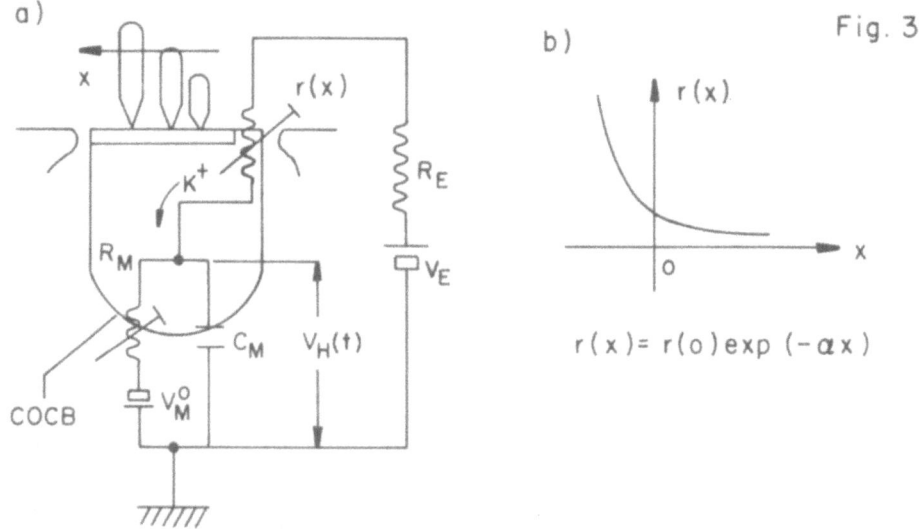

$$r(x) = r(o) \exp(-\alpha x)$$

Fig. 3. Hair cell model. Part (a): schematics; part (b): the nonlinear dependence of the apical-end resistance on x, the
stereocilia deflection relative to the cuticular plate.
V_E, R_E, V_H, R_M and C_M and V_M^o are respectively the endocochlear potential, the endocochlear resistance, the hair cell
membrane potential, the distributed membrane resistance and
capacitance, and the "membrane battery" voltage associated
with the Na^+-K^+ pump.

cross-bridging of the actin fibers and hence the effective stiffness of the
stereocilia. We therefore make the economical hypothesis that the feedback
force, $\Delta f_s(t)$, is proportional to $\Delta V_H(t-\tau)$, where τ denotes a possible time delay between the electrical voltage increment and the resultant mechanical
change.

In order to easily see some of the essential implications of the model, we as-
sume that the frequencies of motion are such that $\omega\tilde{\tau} < 1$ and $\omega\tau_M \ll 1$, where
$\tau_M = R_M C_M$. Then the derivative term in the equation for $\Delta V_H(t)$ may be dropped,
$\Delta V_H(t)$ may be expressed as a simple function of $x(t)$, and Δf_s may be approxi-
mately written as $A \Delta V_H(x) - \tau \frac{\partial}{\partial x} \Delta V_H(x) x$. Substitution of Δf_s in the dynamical equations for ξ_{BM} ($\equiv x/g$) then yields the effective nonlinear BM stiffness
[K(x)] and g resistance [R(x)] functions:

$$K(x) = \frac{k_b}{g} + K_o - \frac{A}{x} \Delta V_H(x), \quad R(x) = \frac{r_b}{g} + \tau A \frac{\partial}{\partial x} \Delta V_H(x) . \qquad (1)$$

k_b and r_b are respectively the intrinsic BM stiffness and resistance parameters. Using a power series for $\Delta V_H(x)$, we find that there is a possibility

of self-sustained oscillations (i.e. spontaneous emissions), Kemp echoes, and sharp BM tuning if

$$\frac{r_b}{g} + \frac{\tau A \alpha r(0) R_M (V_E - V_M^O)}{[R_M + R_E + r(0)]^2} \stackrel{\sim}{<} 0 \ , \tag{2}$$

which requires that A be negative. If this is the case, and if $R_m + R_E \stackrel{\sim}{<} r(0)$, then the coefficients of x^2 in R(x) and K(x) are positive and negative respectively.

4. SOME IMPLICATIONS OF THE FEEDBACK MODEL

The form of R(x) provides for the possibility of stable limit cycles and Kemp echoes. The echoes may be quenched via efferent COCB stimulation which decreases R_M. [Indeed spontaneous cochlear oscillations may arise as a result of a pathology in the efferent pathway which renders R_M large enough to satisfy Eq.(12).] If the feedback process involves some type of actin-myosin interaction which is mediated e.g. by a modulated Ca^{++} influx, then it is plausible that a local depletion of Ca^{++} may result in a reduction in magnitude of the electromechanical coupling parameter A. Zurek (1981) reported fluctuating spontaneous emissions which may possibly be interpreted according to this picture. The nonlinear form of R(x) will give essentially the same type of results (external-tone suppression effects, etc.) as the ad hoc model.

The nonlinear form of K(x) gives qualitatively the phase-intensity results for the psychophysical cubic difference tone obtained with the model of Furst and Goldstein (1982).

The model provides a unified description of the effects of COCB stimulation and EP changes on the parameters of the nonlinear feedback force, and hence on the level of distortion products in the ear canal.

In calculations with $V_M^O \stackrel{\sim}{=} -220$ mV, $V_E \stackrel{\sim}{=} 100$ mV, $r(0) \stackrel{\sim}{=} 200$ MΩ, $R_M \stackrel{\sim}{=} 40$-100 M$\Omega$, and $R_M C_M \stackrel{\sim}{=} 0.16$ ms, we find possible spontaneous oscillation frequencies up to about 5 kHZ (the typical experimental range). The value of $R_M C_M$ was chosen so as to give a cut-off frequency of about 1000 Hz for $\Delta V_H / V_H^O$ (Russell and Sellick, 1978).

In summary, the proposed feedback model accounts, at least qualitatively, for a variety of cochlear acoustic and electromechanical response data. Its basis must of course be challenged and tested by further experimental data on hair cell electromechanics and physiology.

Acknowledgements

This work was supported by the U.S. National Science Foundation.

REFERENCES

Davis, H. (1965). A model for transducer action in the cochlea, *Cold Spring Harbor Symp. Quant. Biol.* 30, 181-189.

Flock, A., and Cheung, H.C. (1977). Actin filaments in sensory hairs of inner ear receptor cells, *J. Cell. Biol.* 75, 339-343.

Furst, M., and Goldstein, J.L. (1982). A cochlear nonlinear transmission-line model compatible with combination tone psychophysics, *J. Acoust. Soc. Am.* 72, 717-726.

Gold, T. (1948). Hearing II. The physical basis of the action of the cochlea, *Proc. R. Soc.* B135, 492-498.

Hall, J.L. (1974). Two-tone distortion products in a nonlinear model of the basilar membrane, *J. Acoust. Soc. Am.* 56, 1818-1828.

Hubbard, A.E., and Geisler, C.D. (1972). A hybrid-computer model of the cochlear partition, *J. Acoust. Soc. Am.* 51, 1895-1903.

Hudspeth, A.J., and Corey, D.P. (1977). Sensitivity, polarity, and conductance change in the response of vertebrate hair cells to controlled mechanical stimuli, *Proc. Natl. Acad. Sci. U.S.A.* 74, 2407-2411.

Kemp, D.T. (1978). Stimulated acoustic emissions from the human auditory system, *J. Acoust. Soc. Am.* 64, 1386-1391.

Khanna, S.M., and Leonard, D.G.B. (1982). Basilar membrane tuning in the cat cochlea, *Science* 215, 305-306.

Kim, D.O., Molnar, C.E., and Pfeiffer, R.R. (1973). A system of nonlinear differential equations modeling basilar-membrane motion, *J. Acoust. Soc. Am.* 54, 1517-1529.

Klinke, R., and Galley, N. (1974). Efferent innervation of vestibular and auditory receptors, *Physiol. Revs.* 54, 316-357.

Koshigoe, S. and Tubis, A. (1982). Unified computational model of stimulated and spontaneous acoustic emissions in the external auditory meatus, *J. Acoust. Soc. Am.* Suppl. 1 71, S17.

Koshigoe, S., Kwok, W., and Tubis, A. (1982). A nonlinear feedback model for mechanical response of outer-hair-cell stereocilia, *J. Acoust. Soc. Am.* Suppl. 1 72, S47.

Koshigoe, S., and Tubis, A. (1983). Frequency domain investigations of cochlear stability in the presence of active elements, to be published in *J. Acoust. Soc. Am.*

Lim, D.J. (1980). Cochlear anatomy related to cochlear mechanics. A review, *J. Acoust. Soc. Am.* 67, 1686-1695.

Mountain, D.C. (1980). Changes in endolymphatic potential and crossed olivocochlear bundle stimulation alter cochlear mechanics, *Science* 210, 71-72.

Mountain, D.C., and Hubbard, A.E. (1982). Injection of the ac current into scala media alters the sound pressure at the tympanic membrane: variations with electrical stimulus parameters, *J. Acoust. Soc. Am.* Suppl. 1 71, S100.

Neely, S.T. (1981). Fourth-Order Partition Dynamics for a Two-Dimensional Model of the Cochlea, doctoral dissertation, Washington University, St. Louis, Missouri.

Neely, S.T., and Kim, D.O. (1983). An active cochlear model showing sharp tuning and high sensitivity, *Hearing Res.* 9, 123-130.

Rhode, W.S., and Geisler, C.D. (1967). Model of the displacement between opposing points on the tectorial membrane and reticular lamina, *J. Acoust. Soc. Am.* 42, 185-190.

Russell, I.J., and Sellick, P.M. (1978). Intracellular studies of hair cells in the mammalian cochlea, *J. Physiol.* 284, 261-290.

Sellick, P.M., Patuzzi, R. and Johnstone, B.M. (1982). Measurement of basilar membrane motion in the guinea pig using the Mössbauer technique, *J. Acoust. Soc. Am.* 72, 131-141.

Siegel, J.H., and Kim, D.O. (1982). Efferent neural control of cochlear mechanics? Olivocochlear bundle stimulation affects cochlear biomechanical nonlinearity, *Hear Res.* 6, 171-182.

Tilney, L.G., DeRosier, D.J., and Mulroy, M.J. (1980). The organization of actin filaments in the stereocilia of cochlear hair cells, *J. Cell. Biol.* 86, 244-259.

Tilney, L.G., Egleman, E., and DeRosier, D.J., and Saunders, J.C. (1983). Actin filaments, stereocilia and hair cells of the bird cochlea. II. The packing of actin filaments in the stereocilia and in the cuticular plate and what happens to the organization when the stereocilia are bent, preprint, University of Pennsylvania.

Weiss, T.F., Mulroy, M.J., and Altmann, D.W. (1974). Intracellular responses to acoustic clicks in the inner ear of the alligator lizard, *J. Acoust. Soc. Am.* 55, 606-619.

Weiss, T.F. (1982). Bidirectional transduction in vertebrate hair cells: A mechanism for coupling mechanical and electrical process, *Hear Res.* 7, 353-360.

Wiederhold, M.L. (1967). A study of efferent inhibition of auditory nerve activity, doctoral dissertation, M.I.T.,Cambridge, MA.

Wilson, J.P., and Sutton, G.J. (1981). Acoustic correlates of tonal tinnitus. In: *Tinnitus*, edited by D. Evered, and G. Lawrenson, (Pitman Medical, London), Ciba Found. Symp. 85.

Wit, H.P., Langevoort, J.C., and Ritsma, R.J. (1981). Freuqency spectra of cochlear acoustic emissions (Kemp-echoes), *J. Acoust. Soc. Am.* 70, 437-445.

Zurek, P.M. (1981). Spontaneous narrowband acoustic signals emitted by human ears, *J. Acoust. Soc. Am.* 69, 514-523.

WAVE REFLECTION IN PASSIVE AND ACTIVE COCHLEA MODELS

E. de Boer

Auditory Physics, ENT Dept. (KNO)
Academic Medical Centre
Amsterdam, the Netherlands

ABSTRACT

In the formulation of a cochlea model, 'waves-to-the-right' (i.e. from stapes to helicotrema) are generally treated as equivalent to 'waves-to-the-left'. In the resonance peak region of the model the conditions for wave propagation vary so rapidly that there is every reason for waves to be reflected. Yet very little evidence of reflected waves is observed, even in an active model. This property is studied by considering various models equipped with a perfectly absorbing wall at the stapes location so that reflected waves can't make the model unstable. A short-wave model does not give rise to reflections. However, a model in which short and long waves are possible shows a preference for waves in the 'normal' direction of propagation (waves-to-the-right), these undergo little or no reflection. Waves in the opposite direction may be reflected when they enter the long-wave region from the short-wave region. The same model is also used to study which degree of impedance irregularity may cause a reflection (an evoked acoustic emission). This degree turns out to be extremely small in an active model: an irregularity of 0.5 per cent extending over the width of two hair cells is sufficient to cause a reflection with the same intensity as the incident wave.

1. INTRODUCTION - STATEMENT OF PROBLEM

Solution methods for mathematical models of the cochlea have evolved a great deal in the past few years. Digital as well as semi-analytical solutions have been obtained for a wide variety of model structures including nonlinear and active models. Not all these solutions, however, provide really good insight into what is going on physically, and this is a serious shortcoming especially in view of problems associated with active behaviour and nonlinearity.

In a linear passive model of the usual type waves do not appear to be reflected at all by the inhomogeneity they encounter, see for the long-wave case de Boer and MacKay (1980) and for the general (passive) case de Boer and van Bienema (1982). Absence of reflection is also found in an active model (de Boer, 1983a). (This property holds true as long as the waves do not reach the helicotrema with an appreciable amplitude - we will tacitly assume this condition to be met throughout this paper).

In more than one sense this property is remarkable (for a more detailed account see de Boer and Viergever, ---). Nonlinear effects create a local disturbance, particularly in the resonance region where these effects are the most pronounced. Doesn't this type of local disturbance produce reflected waves? More serious is the active case: here the locally created disturbance increases the incident wave energy by a large factor (cf. de Boer, 1983b). Doesn't that local disturbance give rise to waves to the left - reflected waves - as well as to the right? And don't these endanger stability? And how about local irregularities in the cochlea as thought responsible for acoustic emissions (cf. Kemp, 1980)? In this paper we will try to answer such questions by studying the destiny of locally created waves in the cochlea. To achieve this, we will study a model in which sound energy can be injected at any desired location. The model is formulated in such a way that eventual creation of reflected waves does not lead to instability, this is done by making the stapes a perfect absorber.

2. ELEMENTS OF THE MODEL - SHORT-WAVE CASE

The basis of the (linear) model considered is a structure consisting of two elongated columns of fluid separated along their lengths by a partition of which a part is flexible. The flexible part of the partition, referred to as the 'basilar membrane' (BM) but actually comprising also the cells and structures of the Organ of Corti, is to be described by way of its mechanical impedance $Z(x)$, where x denotes the location along the length of the BM. This model is, of course, a simplification of reality but no really novel features are discovered when a more complete representation is chosen.

For the formulation of the (three-dimensional) model equations we refer to de Boer (1981). In passive as well as active versions of this model the response in the region of the resonance peak is dominated by 'short waves'. Since this is the main region where reflections may arise because of inhomogeneity we will first consider only short waves. A wave travelling in the direction of positive x in a short-wave cochlea model can be found as the solution to a simple first-order differential equation:

$$\frac{dp(x)}{dx} + g(x)\, p(x) = 0 \tag{1}$$

where p(x) is the pressure and g(x) stands short for $2\omega\rho/Z(x)$ (ω is 2π times
the frequency and ρ is the fluid density). Equation (1) can be derived from
the model equation, Eq. (28) in de Boer (1981), by using the short-wave approximation $Q(k) = 1/k$ for uni-directional waves (k is the wavenumber here). The
solution to Eq. (1) is straightforward, it is a wave of which the (local) wavenumber is equal to $\{-ig(x)\}$. An actual short-wave model should equally well
accomodate waves going to the left as to the right. Thus there should exist
a second-order differential equation of which the two independent solutions
are: the solution to Eq. (1) and the solution to the same equation with a mi-
nus instead of a plus sign. This equation turns out to be:

$$\frac{d^2p(x)}{dx^2} - \frac{g'}{g}\frac{dp(x)}{dx} - g^2p(x) = 0 \qquad (2)$$

where g stands short for g(x) and g' for its first derivative.

There are, of course, two boundary conditions to be met, one at the stapes
and one at the helicotrema. The latter one prescribes, as usual, the wave
to go to zero amplitude. The other boundary condition will be formulated in
an unusual way, namely by stating that any wave travelling to the left is
completely absorbed by the stapes. There is then no danger that a left-going
wave, after being reflected by the stapes, can cause instability of the model.
Acoustic energy can now be injected at any desired location x. To achieve
freedom of reflection at the stapes we proceed as follows. Equation (2) is
manipulated so that its second term is zero. The resulting equation is recognized to describe wave propagation in a non-homogeneous electrical transmission line. (It is also similar in form to the wave equation for a long-wave
cochlear model but that is of no concern here). The electrical transmission
line can be terminated at one of its ends by its characteristic impedance
thus ensuring that no reflection can take place at that end. This is implemented at the stapes end; for the short-wave case the characteristic impedance
turns out to be constant, independent of Z(x) (as has also been found by
Zwislocki, 1983, in a completely different context).

3. SHORT WAVES AND THEIR REFLECTION

In view of space limitations we must refrain from showing results for the
short-wave active case in detail. We can describe these briefly in the following terms. When sound energy is injected at a point inside the region of

the response maximum, waves emerge from that location travelling in both directions. Because the BM is active over the larger part of this region, both
waves are amplified as they travel along. The left-going wave appears to be
reflected somewhere, and the reflected wave, now travelling to the right, is
amplified further. The latter wave eventually overrides the former one so that
a dominant right-going wave travels toward the injection point and beyond it.
It should be well remembered that this can only be true in a locally active
structure (in contrast to the 'paradoxical wave travel' problem considered
by Bekesy, 1960).

Is this now an example of reflection of left-going waves that might be expected on theoretical grounds (de Boer and Viergever, ---)? Some simple ma-
nipulations of the model show that this is not so. For instance, displacement
of the left-hand end (stapes location) produces a variation of the reflection
pattern. Apparently, the reflection is an artefact, caused by imperfect absorption of energy by the stapes. In turn, this is due to imperfect digital
implementation of the no-reflection condition in the computer program. The
short-wave model appears to be sensitive to this type of imperfection.

The interpretation of these results is that in the short-wave case waves will
not be reflected - no matter how inhomogeneous the medium of propagation is.
This property is only subject to the condition that $Z(x)$ is of such a form
that the solution to eq. (1) has a one-sided spectrum (de Boer and Viergever,
---).

4. SHORT AND LONG WAVES

The results presented in the preceding section were essentially negative. In
the region of the cochlea where short waves dominate no clear tendency for
(internal) reflection exists, neither for waves going to the right nor for
those to the left. We know from a previous study (de Boer and MacKay (1980))
that the case for long waves is similar. It remains to study the transition
from long to short waves and vice versa. We need here a computation technique
which applies to a general type of model, i.e., one that incorporates short
as well as long waves, and that treats left- and right-going waves in exactly
the same way. We again start from the formulation in de Boer (1981). For the
purpose of this paper an even function of k is needed to approximate the ker-
nel function $Q(k)$ (cf. de Boer and van Bienema, 1982). We use the following

rational function:

$$Q(k) = \frac{c_0}{k^2} \frac{1 + a_2 k^2}{1 + a_4 k^2} \tag{3}$$

where the constants c_0, a_2 and a_4 are chosen to obtain the best approximation
to the $Q(k)$ function for a 3-dimensional model. With substitution of eq. (3)
the integral equation for the model can easily be converted into a differential equation of the fourth order. Setting up this equation is simple and
straightforward. As in the preceding case, the left-hand end is terminated in
its characteristic impedance - this time evaluated for long waves.

For the purpose of clarity real and imaginary components of the impedance $Z(x)$
are chosen more or less independently in this paper. We keep, however, the
function $Z(x)$ close to realizability by verifying that the Fourier transform
$Y(k)$ of $1/Z(x)$ is approximately a one-sided function of the wavenumber k
(cf. de Boer and Viergever, ---). This is also important for active impedances
as considered here. The form of $Z(x)$ is the same as the one used in de Boer
(1983a) but with the following parameters: $\varepsilon = 1$, $a_0 = 0.1$, $a_1 = 2.5 + 3i$,
$x_{range} = 0.2$, all in cgs units. The point $x = 0$ corresponds to the 'resonance
location' where Im $\{Z(x)\}$ is zero. Fig. 1 shows the shape of the $Z(x)$ function.

Fig. 2 shows results of computations with this method, velocity of the BM
against location x. The constant c_0 is chosen so as to have correct long-
wave behaviour in a model where the fraction $\varepsilon = 0.1$ of the partition width
is occupied by the BM. The other constants are chosen such as to have approximately the same short-wave behaviour as in the preceding section. Fig. 2
contains four curves. Curves A_1 (amplitude) and A_2 (phase) pertain to the

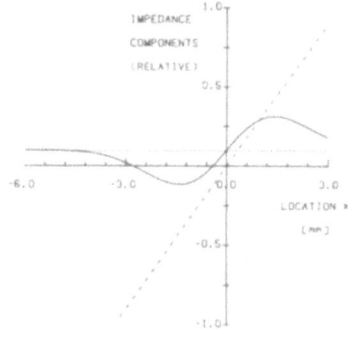

Fig. 1. The impedance function $Z(x)$.
Solid line: real part,
dashed line: imaginary part.
Formula used is

$$\frac{Z(x)}{2\omega\rho\varepsilon m_0} = a_0 + a_1 x \; exp\{-4(\frac{x}{x_{range}})^2\}$$

with $m_0 = 0.05$.

'normal' situation: injection occurs near the left-hand border
(at x = -5.7 $[mm]$). These curves show the familiar response peak which suggests an internal amplification on the order of 40 dB. More important are the
curves B_1 and B_2 which show the situation where injection occurs inside the
active region (at x = -0.3 $[mm]$). *A part of the wave is seen to travel
toward the injection point.* In fact, the situation can be understood in the
same way as before: at the injection site two waves originate, one going to
the right and one to the left. On their paths in the active region both waves
are amplified. The left-going wave is reflected somewhere at the left-hand
side of the peak and it is again amplified while travelling along after this
reflection. This amplification eventually results in a dominating right-going
wave approaching the site of energy injection.

The problem is, again: what causes the reflection of the left-going wave?
From various experiments it can provisionally be concluded that the reflection
is not due to imperfection at the energy-absorbing wall. It is not due to the
transition between active and passive behaviour either. The reflection appears
to be associated with the transition between short and long waves, and as such
it seems to be an intrinsic feature of this type of model. Further study is
needed to elucidate this property.

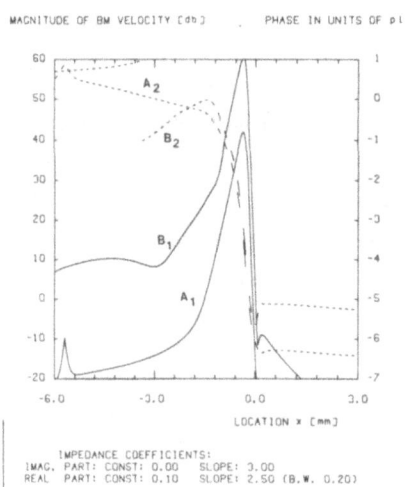

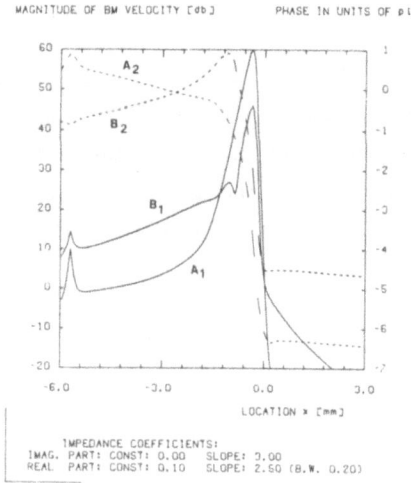

*Fig. 2. Injection at two points
(see text).
(for left-going wave, phase is shown
modulo 2π).*

Fig. 3. Reflection by an irregularity.

5. LOCAL REFLECTION - EVOKED EMISSIONS

With the experience acquired it is now possible to interpret the effect of a local disturbance in propagation conditions. Fig. 3 presents results of computations. Curves A_1 and A_2 show, just as in Fig. 2, the 'normal' wave pattern resulting from injection at $x = -5.7$ [mm]. For reasons of clarity the response has been shifted upward by 20 dB. Curves B_1 and B_2 show the response from driving the model in the same way but using an impedance function that has an irregularity. In fact, the impedance has been changed by 5 per cent from $x = -0.3$ to $x = -0.276$ [mm]. The curves are seen to be most similar to the B-curves in Fig. 2. A very small irregularity in $Z(x)$ apparently sets up a reflected wave that fully dominates the wave pattern. In effect, the initial part of the wave near the stapes is completely swamped by a left-going wave, and the magnitude of the remaining part has been altered considerably. This inversion of wave direction occurs with an irregularity as small as 0.5 per cent over this region (24 µm long, i.e., extending over 2 hair cells).

The effect of irregularities is largest in the region of the peak and slightly to the right of the top. This is easily understood: it is here where the reflected wave is amplified the most. That the active model is so sensitive to local irregularities is also easily understood: on the entire pathway from stapes to reflection site and back to the stapes, waves undergo an amplification that is twice that of the incident wave alone (i.e., about 80 dB!). Note, incidentally, that the total propagation delay is also doubled.

We should stress, finally, that the phenomenon observed in Fig. 3 cannot be seen in a more realistic model where the stapes is not a perfect absorber: on the basis of thermal noise such a model would go into spontaneous oscillation. The expedient of a perfectly absorbing stapes in an active model of the cochlea appears to be very useful indeed: it allows us to understand the physics of the situation better. On the other hand, the fact that we have to use such a resource also shows how little we really know of what is going on in this type of active structure.

6. CONCLUDING REMARKS

The following, partially tentative, conclusions may be drawn from this work:

1) waves going to the right are not reflected, no matter how inhomogeneous the medium is - this is subject to the condition mentioned in section 3.

2) waves going to the left may be reflected in the transition between the short- and long-wave regions.

3) an active model is extremely sensitive to slight irregularities of wave propagation conditions.

4) when an evoked emission occurs from an irregularity, its latency is approximately twice that of the original incident wave to the site of the irregularity.

Only a few of the questions posed in the Introduction have received an answer. For instance, it is not clear yet why in an active system all left-going waves tend to annihilate each other. Are these really being completely dominated by right-going waves? Or do all microscopically created left-going waves tend to cancel? The simple 'two-way' short-wave model developed in section 2 may be a convenient vehicle for a semi-analytical study of these problems. More difficult is the topic of reflection between the long- and short-wave regions. Many problems of active systems thus remain somewhat obscure: it may take a while before further intricacies - such as those contributed by nonlinearity, for instance - can be grasped completely.

Acknowledgement

This study was supported by the Netherlands Foundation for Pure Research.

REFERENCES

von Bekesy, G. (1960). *Experiments in hearing* (McGraw-Hill, New York).

de Boer, E. (1981). Short waves in three-dimensional cochlea models: Solution for a "Block" model, *Hearing Research* 4, 53-77.

de Boer, E. (1983a). No sharpening? A challenge for cochlear mechanics, *J. Acoust. Soc. Am.* 73, 567-573.

de Boer, E. (1983b). Power amplification in an active model of the cochlea - short-wave case, *J. Acoust. Soc. Am.* 73, 577-579.

de Boer, E. and MacKay, R. (1980). Reflections on reflections, *J. Acoust. Soc. Am.* 67, 882-890.

de Boer, E. and van Bienema, E. (1982). Solving cochlear mechanics problems with higher-order differential equations. *J. Acoust. Soc. Am.* 72, 1427-1434.

de Boer, E. and Viergever, M.A. (--). Wave propagation and dispersion in the cochlea, *Hearing Research* (submitted for publication).

Zwislocki, J.J. (1983). Sharp vibration maximum in the cochlea without wave reflection, *Hearing Research* 9, 103-112.

MODELLING AN ACTIVE, NONLINEAR COCHLEA

S.M.van Netten, H.Duifhuis

Dept. of Biophysics, Rijksuniversiteit Groningen
The Netherlands

ABSTRACT

The current interpretations of data pertaining to cochlear mechanics appear to converge on the points that: 1. the ratio between pressure across the basilar membrane and membrane velocity is nonlinear, and 2. the mechano-electrical transduction process may be reversible, so that it can operate as an active mechanical process. It is becoming more widely accepted that these two properties originate in the hair cells. It is quite conceivable that there are small unsystematic as well as systematic differences between mechanical properties of different hair cells. In order to analyse the behaviour of an active nonlinear cochlea we started to study a one-dimensional description of the cochlear hydromechanics, employing the usual equations for pressure and velocity. Instead of defining a local impedance Z(x,ω) as the pressure-to-velocity-ratio in the linear case, where it describes a linear second order oscillator, we now use a form describing a Van der Pol-oscillator. Evaluation for a single tone stimulus shows relatively ·harp excitation patterns at low intensities. With increasing intensity the p·.aks broaden and shift basalwards. The properties of the Van der Pol-oscillι.tor, which are extensively documented, appear to provide a fruitful analysis tool, which can, a.o. readily be extended to a multi-tone stimulus.

1. INTRODUCTION

Theoretical cochlear mechanics has received considerable renewed interest over the past decade, as is reflected in the contributions to this symposium. The first impulse for the revival stemmed from the new basilar membrane data obtained around 1970 (Johnstone et al., 1970; Rhode, 1971; Wilson and Johnstone, 1972) showing sharper frequency tuning than assumed so far. The second came from the availability of more powerful tools (hardware and software) for evaluating more realistic models. It became feasible to evaluate 2- and 3-dimensional models, in addition to the 1-dimensional model (e.g. Lesser & Berkley, 1972; Allen, 1977; Steele & Taber, 1979; De Boer, 1980a). Even the 3-dimensional models contain many simplifications of the physical reality. Most of these are sufficiently justifiable (Viergever, 1980). Many theorists, however, realize that one can not describe the ratio of cochlear partition velocity and pressure difference adequately with a linear, passive, second order impedance function. There is now an abundance of evidence that the intact cochlea behaves nonlinearly. This point has been pursued in several studies (e.g., Hubbard and Geisler, 1972; Hall, 1974; Matthews, 1980).Evidence that the cochlea may be mechanically active, at least at some frequencies, is compiling steadily (e.g., Kemp, 1978) and theories accounting for this are being developed (e.g., Kim et al., 1980). And thirdly, several investigators are beginning to take the microstructure of the cochlea into account (e.g., Duifhuis and Van de Vorst, 1980; Allen, 1980; Zwislocki and Kletsky, 1980). Inevitably this leads to higher order mechanical models. In addition to this, new developments in experimental cochlear mechanics revealed still sharper tuning properties (Khanna and Leonard, 1982; Sellick et al., 1982). These recent data cannot be accounted for with passive, linear models (de Boer, 1983).

At this point we believe that it is possible and desirable to make significant progress with a macro-description of the cochlear partition by appropriate modification of the local impedance parameters. Thus we will not discuss the micromechanical structure and we will not specify the sources of nonlinear and active behaviour (which we assume to be the hair cells). We will explore, however, the combined effects of nonlinearity and activity.

2. MODELLING ACTIVE AND NONLINEAR BEHAVIOUR

As a starting point in the analysis of the combined effects of nonlinearity and active behaviour in cochlear hydromechanics we analyse the 1-dimensional long-wave model (Zwislocki, 1950). For the displacement y of the basilar membrane at location x we have, as usual

$$\left[m(x)\frac{\partial^2}{\partial t^2} + r(x)\frac{\partial}{\partial t} + c(x)\right]y = p(x)\ \sin(\omega t + \phi(x)) \tag{1}$$

where $m(x)$ is relevant mass per square meter (kg/m^2), $c(x)$ is relevant stiffness per square meter (Pa/m), $p(x)\ \sin(\omega t + \phi(x))$ is the pressure difference across the membrane at x (Pa), but where active and nonlinear behaviour are incorporated in the damping term

$$r(x) = -\ R_1(x) + R_2(x)\ \frac{\partial^2 y}{\partial t^2} \quad Pa.s/m \tag{2}$$

with $R_1(x)$ and $R_2(x) > 0$. The choice of the nonlinear term is not entirely original (see, e.g., Hall, 1974). It adequately describes compression or saturation at high stimulus levels. Note that the active behaviour can only be pronounced at low input levels, i.e., when $r(x)$ is small.

Differentiating Eq. (1) with respect to t and substituting

$$\varepsilon = \sqrt{\frac{R_1^2}{mc}}\ ,\quad v = \sqrt{\frac{3R_2}{R_1}}\ \frac{\partial y}{\partial t}\ ,\quad T = \omega_r t,\quad \Omega = \frac{\omega}{\omega_r}\ (\text{with } \omega_r = \sqrt{\frac{c}{m}}),\ \text{and } q = p\sqrt{\frac{3R_2}{R_1}}\ \frac{\omega}{c}$$

we arrive at the Van der Pol-equation

$$\ddot{v} - \varepsilon \, (1 - v^2) \, \dot{v} + v = q \, \cos(\Omega T + \phi) \tag{3}$$

where the dots indicate differentiation with respect to T and where we have
omitted the explicit indication of the dependence on x. Note that $\Omega T = \omega T$. The
applicability of the Van der Pol-equation in this context was first proposed
by Johannesma (1980).

Properties of the Van der Pol-equation have been discussed extensively in the
literature (e.g., Van der Pol, 1927; Stoker, 1950; McLachlan, 1950; Nayfeh and
Mook, 1979).

We recall some characteristics for the free and the driven oscillator.

1) The free oscillator, q = 0, generates a periodic waveform with a fundamental of approximately $(1 - \varepsilon^2/16)$. The ultimate stable oscillation is independent of the system's boundary conditions. The dimensionless factor ε (the
magnitude of which is comparable with the damping factor δ in the passive
linear models) determines the specific behaviour of the solution of Eq. (3).
Hence, qualitatively the solution is independent of the strength of the nonlinear term R_2. If $\varepsilon < 1$, then the higher harmonics in the solution may be
neglected, yielding (e.g., McLachlan, 1950)

$$v(T) = v_o \, \exp \, (\varepsilon T/2) \, \cos T \, \left[1 + \tfrac{1}{4} \, v_o^2 \, (\exp \, (\varepsilon T) - 1)\right]^{-\frac{1}{2}} \tag{4}$$

where v_o is the initial (dimensionless) velocity amplitude $v(o)$. If one assumes that the energy source $(-R_1)$ can be switched on, or switched off, then
the risetime and decaytime can be evaluated. For the 90% risetime one ob-
taines

$$\tau_{0,9}^{+} \approx \frac{1}{\varepsilon \omega_r} \, (2,8 - 2 \, {}^e\!\log v_o) \tag{5}$$

which depends on the initial velocity v_o, and for the 50% decaytime (half-
live time)

$$\tau_{0,5}^{-} \approx \frac{3}{\varepsilon \omega_r} \, . \tag{6}$$

Even though nonlinear damping causes the decay, the decaytime is independent
of R_2.

2) For the sinusoidally driven oscillator, $q \neq 0$, the response depends on

driving frequency and pressure magnitude. Either an oscillation developes only
at the driving frequency, or one obtains a so-called combination oscillation
consisting of fundamental and harmonics of the driving frequency and a free-
oscillation component. If $\varepsilon < 1$, then the harmonics can be neglected in first
approximation (Van der Pol, 1927). Thus, if $\varepsilon \ll 1$ and if driving frequency
and pressure are chosen properly, then an oscillation builds up at the driving
frequency only. This phenomenon is known as entrainment. The dependence of v on
on q and ω is implicitely given in

$$F^2 = \rho \left[\sigma^2 + (1 - \rho)^2 \right] \tag{7}$$

with $F = \dfrac{q}{2\varepsilon\omega}$, $\sigma = \dfrac{1-\Omega^2}{\varepsilon\Omega}$, and $\rho = \tfrac{1}{4} v^2$ (e.g., Stoker, 1950).

If the driving term contains two frequencies, then Eq. (3) becomes

$$\ddot{v} - \varepsilon(1-v^2)\, \dot{v} + v = q_1\cos(\Omega_1 T + \phi_1) + q_2\cos(\Omega_2 T + \phi_2). \tag{8}$$

This case has hardly been studied in the literature. It is of course of prime
interest in this context in view of combination tone - and lateral suppression
behaviour of the cochlea. Extending Van der Pol's method straightforwardly we
obtained the coupled equations

$$F_1^2 = \rho_1 \left[\sigma_1^2 + (1 - \rho_1 - 2\rho_2)^2 \right]$$
$$F_1^2 = \rho_2 \left[\sigma_2^2 + (1 - \rho_2 - 2\rho_1)^2 \right] \tag{9}$$

where F_1, σ_1 and ρ_1 are defined as above. For the single Van der Pol-oscillator this solution produces already some suppression. This is shown in Fig.1,

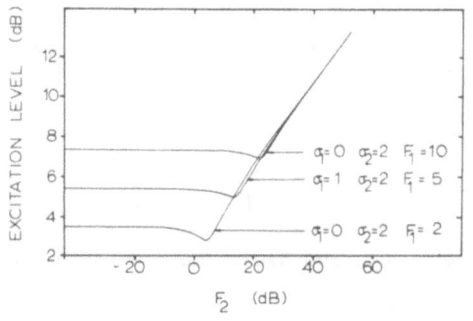

Fig.1. Two-tone suppression in a single Van der Pol-oscillator.

which gives the response measure $\rho_1 + \rho_2$ as a function of the strength of component 2, proportional to F_2, with F_1, σ_1 and σ_2 as parameters. (We stress that this is the result for a single oscillator. It is not a quantitative prediction of cochlear lateral suppression, which is the combined result of all the fluid-coupled oscillators.) Just as with the linear oscillator we may define the ratio of pressure and velocity as a local impedance. In the dimensionless normalized units $\zeta = F/\sqrt{\rho}$, or for a multicomponent driving pressure, generalizing Eq. (9):

$$\zeta_\ell = \frac{F\ell}{\sqrt{\rho_\ell}} = - i \, \sigma_\ell - (1 - \rho_\ell - 2 \sum_{k \neq \ell} \rho_k) \qquad \ell = 1, \ldots L \qquad (10)$$

with L the number of components. Inspection of Eq. (10) reveals that the nonlinear interaction of tones affects the real parts of ζ_ℓ only and that the interaction term is independent of phase.

3. MODELLING THE COCHLEA

The behaviour of a one-dimensional cochlea is, for the long-wave approximation, described by the second order differential equation

$$\psi_{xx} - \frac{2i\omega\rho}{hZ(x,\omega)} \, \psi = 0 \qquad (11)$$

with the appropriate boundary conditions. Equation (11) follows the notation used by De Boer (1980 b); henceforth ρ denotes fluid density. The wave equation is equivalent with the definition of the cochlear (point) impedance

$$Z(x,\omega) = - \frac{p(x,\omega)}{w(x,\omega)} \qquad (12)$$

where $p(x,\omega)$ is the pressure difference across the cochlear partition and $w(x,\omega)$ its velocity, because

$$w(x,\omega) = - \frac{1}{b} \, \psi_{xx} \qquad (13)$$

and $\qquad p(x,\omega) = \frac{2i\omega\rho}{b \, h} \, \psi. \qquad (14)$

Substituting Eq. (10), after appropriate scaling, into Eq. (11), which amounts to implying that Eqs. (13) and (14) remain valid, provides the following set of coupled nonlinear second order differential equations:

$$2 \sum_{k \neq \ell} \left| \psi_{xx}^{(k)} \right|^2 \psi_{xx}^{(\ell)} + \left| \psi_{xx}^{(\ell)} \right|^2 \psi_{xx}^{(\ell)} + \alpha_\ell \psi_{xx}^{(\ell)} + \beta_\ell \psi^{(\ell)} = 0 \qquad (15)$$

$$\ell = 1 \dots L$$

with

$$\alpha_\ell = \frac{4}{3} \frac{b^2}{R_2} \left[i \left(\omega_\ell m - \frac{c}{\omega_\ell} \right) - R_1 \right]$$

$$\beta_\ell = - \frac{8}{3} \frac{i \omega_\ell \rho}{h} \frac{b^2}{R_2} .$$

The sum term in Eq. (15) will describe suppression effects due to interaction with neighbouring components, whereas the second term will be responsible for saturation or compression at higher levels. The third and fourth term are related directly to Eq. (11). For a single tone stimulus Eq. (15) reduces to

$$\left| \psi_{xx} \right|^2 \psi_{xx} + \alpha \psi_{xx} + \beta \psi = 0. \qquad (15a)$$

Because ε was supposed to be small, neither Eq. (15) nor Eq. (15a) contain solutions for the strength of harmonics or intermodulation frequencies. The latter simplification is probably not justified for the multitone stimulus.

4. NUMERICAL EVALUATION AND DISCUSSION

We have evaluated Eq. (15a), the single tone case, using subroutine DTPTB from the International Mathematical and Statistical Libraries (IMSL). The parameter values used are $m = 0.5$ kg/m^2 and $c = 10^{10} \exp(-300 \ x)$ Pa/m, $b = 10^{-3}$m, $h = 10^{-3}$m, $\rho = 10^3$ kg/m^3 in line with other recent studies. The active "damping" term was tentatively chosen in such a way that $\varepsilon = 0.05$ over the entire cochlea. The nonlinear part of the damping was also chosen arbitrarily, in such a way that R_1/R_2 is independent of x, yielding $R_2 = 10^7. \exp(-150 \ x)$ Pa.s^3/m^3.

Figure 2 shows the magnitude of the basilar membrane velocity for several boundary conditions at the stapes. For comparison a linear velocity-pattern has been added (with $\delta = 0.05$). Figure 3 is a replot of Fig.2, but normalized re input level. The most striking results are the relative sharp peak for low input levels and the broadening and basalward shift of the peaks with increasing levels. The sharper peak can account for the latest data (Khanna and Leonard, 1982; Sellick et al., 1982). At this point we have not attempted a precise parameter fit because the data set is still limited, but note that height and sharpness of the peak are determined by the negative damping

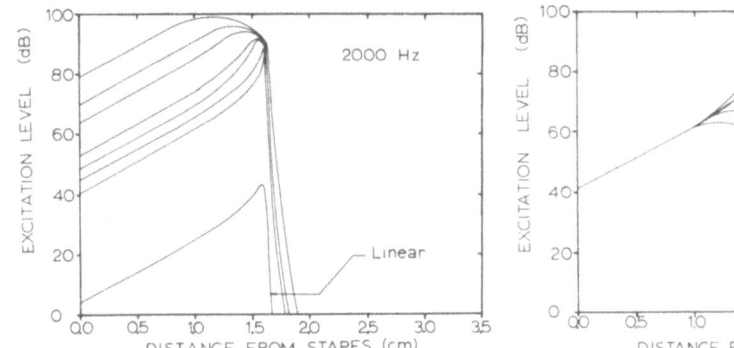 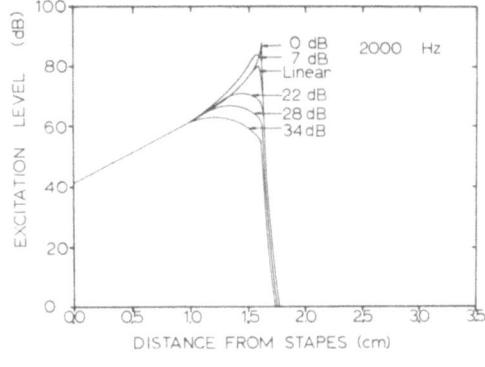

Fig.2. Envelopes of velocity patterns for several input velocity levels at the stapes.

Fig.3. As Fig.2 but now normalized with respect to the stapes velocity. (For the 0 dB pattern the normalized velocity v at the resonance point is 2.3, i.e. sufficiently within the validity of the Van der Pol-oscillator).

parameter. The broadening and basalward shift with increasing level agrees with earlier data (Rhode, 1971) and with results from passive nonlinear models (Hall, 1974), and is obviously determined by the nonlinear damping parameter.

In Fig.4 the phases corresponding with the 28 dB response have been plotted for four frequencies. The dips at the resonance points are not new; however, in this model their sharpness increases with increasing input level. So it appears that, as the velocity excitation pattern broadens at resonance, the phase dip tends to sharpen.

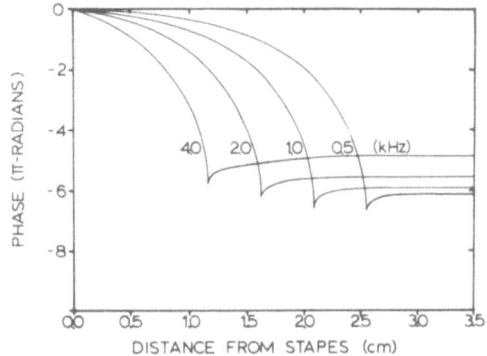

Fig.4. Phase characteristic corresponding with the 28 dB velocity response pattern.

Thus far, phase dips have only appeared in multidimensional studies and their meaning is not yet quite clear. Phase dips can be interpreted as points where two waves, one coming from the stapes and one coming from the apex, vanish. Unfortunately, phase-data beyond the resonance point are hardly available.

150

Summarizing, we state that the nonlinearity and active behaviour of the
Van der Pol-oscillator seem to furnish a promising description of these properties of the cochlear partition. In view of the behaviour of a two-tone
driven single Van der Pol-oscillator, as shown in Fig.1, it seems plausible
that the complete model produces two tone suppression. This expectation can
now be verified through evaluation of Eq. (15) for L = 2. The effects of intermodulation products should then also be taken into accout.

Acknowledgements

*This contribution is supported in part by the Netherlands Organization for
the Advancement of Pure Research (Z.W.O.).*

REFERENCES

Allen, J.B. (1977). Two-dimensional cochlear fluid-model: New results, *J.Acoust. Soc.Am.* 61, 110-119.
Allen, J.B. (1980). A cochlear micromechanic model of transduction. In: *Psychophysical, Physiological and Behavioural Studies in Hearing*, edited by G.van den Brink and F.A.Bilsen (Delft University Press, Delft), pp. 85-93.
De Boer, E. (1980a). A cylindrical cochlea model: The bridge between two and three dimensions, *Hearing Res.* 3, 109-131.
De Boer, E. (1980b). Auditory physics. Physical principles in hearing theory 1, *Physics Rep.* 62, 87-174.
De Boer, E. (1983). On active and passive cochlear models-towards a generalised analysis, submitted for publication.
Duifhuis, H., and Van de Vorst, J.J.W. (1980). Mechanics and nonlinearity of hair cell stimulation, *Hearing Res.* 2, 493-504.
Hall, J.L. (1974). Two-tone distortion products in a nonlinear model of the basilar membrane, *J.Acoust.Soc.Am.* 56, 1818-1828.
Hubbard, A.E., and Geisler, C.D. (1972). A hybrid-computer model of the cochlear partition, *J.Acoust.Soc.Am.* 51, 1895-1903.
Johannesma, P.I.M. (1980). Narrow band filters and active resonators. In: *Psychophysical, Physiological and Behavioural Studies in Hearing*, edited by G.van den Brink and F.A.Bilsen (Delft University Press, Delft), pp. 62-63.
Johnstone, B.M., Taylor, K.J., and Boyle, A.J. (1970). Mechanics in the guinea pig cochlea, *J.Acoust.Soc.Am.* 47, 504-509.
Kemp, D.T. (1978). Stimulated acoustic emission from within the human auditory system, *J.Acoust.Soc.Am.* 64, 1386-1391.
Khanna, S.M., and Leonard, D.G.B. (1982). Basilar membrane tuning in the cat cochlea, *Science* 215, 305-306.
Kim, D.O., Neely, S.T., Molnar, C.E., and Matthews, J.W. (1980). An active cochlear model with negative damping in the partition: Comparison with Rhode's ante- and post-mortem observations. In: *Psychophysical, Physiological and Behavioural Studies in Hearing*, edited by G.van den Brink and F.A.Bilsen (Delft University Press, Delft), pp. 7-14.
Lesser, M.B., and Berkley, D.A. (1972). Fluid mechanics of the cochlea. Part 1, *J.Fluid Mech.* 51, 497-512.
Matthews, J.W. (1980). *Mechanical modelling of nonlinear phenomena observed in the peripheral auditory system*, ScD-thesis, Washington (MD).

McLachlan, N.W. (1950). *Ordinary nonlinear differential equations in engineering and physical sciences* (Oxford University Press, Oxford).

Nayfeh, A., and Mook, D. (1979). *Nonlinear oscillations* (Wiley, New York).

Rhode, W.S. (1971). Observations of the vibration of the basilar membrane in squirrel monkeys using the Mössbauer technique, *J.Acoust.Soc.Am.* 49, 1218-1231.

Sellick, P.M., Patuzzi, R., and Johnstone, B.M. (1982). Measurements of basilar membrane motion in the guinea pig using the Mössbauer technique, *J.Acoust.Soc.Am.* 72, 131-141.

Steele, C.R., and Taber, L.A. (1979). Comparison of W.K.B. and finite difference calculations for a two-dimensional cochlear model, *J.Acoust.Soc.Am.* 65, 1001-1006.

Stoker, J.J. (1950). *Nonlinear vibrations in mechanical and electrical systems*, (Interscience publishers, New York).

Van der Pol, B. (1927). Forced oscillations in a circuit with nonlinear resistance, *Philosophical magazine* 3, 65-80.

Viergever, M.A. (1980). *Mechanics of the inner ear - a mathematical approach*, academic thesis (Delft University of Technology).

Wilson, J.P., and Johnstone, J.R. (1972). Capacitive probe measures of basilar membrane vibration. In: *Hearing Theory*, edited by B.L.Cardozo (I.P.O., Eindhoven), pp. 172-181.

Zwislocki, J.J. (1950). Theory of the acoustical action of the cochlea, *J. Acoust.Soc.Am.* 22, 778-784.

Zwislocki, J.J.,and Kletsky, E.J. (1980). Micromechanics in the theory of cochlear mechanics, *Hearing Res.* 2, 502-512.

NONLINEAR AND ACTIVE MODELLING OF COCHLEAR MECHANICS: A PRECARIOUS AFFAIR

Rob J. Diependaal, Max A. Viergever

Department of Mathematics and Informatics
Delft University of Technology, The Netherlands

ABSTRACT

We have developed a numerical solution method for one-dimensional cochlea models in the time domain. The method has particularly been designed for models with a cochlear partition having nonlinear and active mechanical properties. Our starting point is the partial differential equation describing the response of the basilar membrane to stapes movements. Using Galerkin's principle, we reduce this equation to a system of ordinary differential equations in the time variable. The remaining initial value problem is solved by means of an explicit fourth order Runge-Kutta scheme with a variable-step routine.
The calculations show that the response of an active, nonlinear cochlea model is very much contingent on the solution method. Approaches that exclude or suppress reflection phenomena must be considered inappropriate.
The results obtained so far suggest that nonlinear and active modelling of cochlear mechanics should comply with the following conditions in order to be consistent with experimental data:
(i) The active and nonlinear properties must be tightly linked,
(ii) Omission of these properties must yield the classical description
of cochlear macromechanics,
(iii) The active behaviour of the partition must be localized in a small longi-
tudinal region of the cochlea.

1. INTRODUCTION

The field of cochlear modelling has recently changed its focus considerably.
Stimulated by experimental results concerning cochlear acoustic emissions
(pioneered by Kemp, 1978) many investigators have sought for ways to include
active features in descriptions of cochlear functioning. A new impetus was
given by the ascertainment that active behaviour clearly manifests itself at
the level of basilar membrane (BM) vibration. While the earlier measurements
including those of Rhode (1971, 1978) could all be matched quite well by the
response of a passive model (Viergever and Diependaal, 1983), the results
from Khanna and Leonard (1982) and Sellick, Patuzzi and Johnstone (1982) do not
admit such a match. De Boer (1983a,b) has shown convincingly that models of
cochlear mechanics must be endowed with active properties (for instance, negative BM resistance) in order to produce responses that compare favourably with
the new data.

Active properties have been included in linear cochlea models by Kim, Neely,
Molnar, and Matthews (1980), Neely (1981), De Boer (1983a) and Neely and Kim
(1983). These studies have yielded interesting results, e.g. that locally
active mechanical behaviour of the BM does not preclude an overall stable

response. It is more natural, however, to analyse cochlear activity in a nonlinear context, since active and nonlinear features are either both present
(in an intact cochlea) or both absent (in a damaged cochlea). A first attempt
to combine the two can be found in another contribution to these proceedings
(Van Netten and Duifhuis, 1983).

Inclusion of active properties endangers the stability of the model response.
Therefore, considerable attention should be given to the choice of solution
method: no features may be introduced or obscured by the mathematical treatment of the model equations. Linear active models can relatively easily be
solved in the frequency domain by means of convergent numerical approximations,
if the models count no more than two spatial dimensions. Numerical methods for
nonlinear models are more complex and much more time-consuming, though, because the solution is to be obtained in the time domain. This suggests the use
of asymptotic methods, which are computationally fast and in addition will give
more insight into the pertaining physical mechanisms than numerical techniques
do. Indeed, Van Netten and Duifhuis (1983) utilized a linearizing asymptotic
technique to transform their model equations to the frequency domain.

The aim of the present work is to provide a frame of reference for asymptotic
solution methods. To this end we formulate the cochlear model in the time
domain and solve the equations using a straightforward numerical technique.
For reasons of simplicity we have confined ourselves to a one-dimensional (1D)
treatment. The method can, however, in principle (that is, apart from problems
related to computer storage and computational speed) easily be extended to
2D and 3D models. We consider the method as outlined in the sequel still as
tentative. More model examples than the ones discussed in this paper will have
to be covered before final conclusions about design and treatment of nonlinear
and active models can be deduced.

2. MODEL AND METHOD

The geometry of the one-dimensional cochlear model and the assumptions upon
which the model is based are discussed in Viergever (1980, chapter 2). The
partial differential equation describing the response of the BM to stapes movements is

$$\frac{\partial^2}{\partial x^2} \left\{ m \frac{\partial^2 u}{\partial t^2} + r \frac{\partial u}{\partial t} + su \right\} - \frac{2\rho\beta}{a} \frac{\partial^2 u}{\partial t^2} = 0, \quad 0 < x < L, \quad t > 0, \tag{1}$$

with the initial and boundary conditions

$$u(x,0) = 0 \qquad\qquad , \quad 0 \le x \le L, \qquad\qquad (2)$$

$$\frac{\partial u}{\partial t}(x,0) = 0 \qquad\qquad , \quad 0 \le x \le L, \qquad\qquad (3)$$

$$\frac{\partial}{\partial x}\left(m\,\frac{\partial^2 u}{\partial t^2} + r\,\frac{\partial u}{\partial t} + su\right)\bigg|_{x=0} = f(t) , \qquad t \ge 0 \qquad , \qquad (4)$$

$$\left(m\,\frac{\partial^2 u}{\partial t^2} + r\,\frac{\partial u}{\partial t} + su\right)\bigg|_{x=L} = 0 \qquad , \qquad t \ge 0 \qquad . \qquad (5)$$

The meaning of the symbols is

x	distance along the BM (x=0 at the stapes)
t	time variable
L	length of the BM
a(x)	cross-sectional area of the channels
β(x)	BM width
ρ	density of the cochlear fluids
m(x)	BM mass per unit area
r(x)	BM resistance per unit arèa
s(x)	BM stiffness per unit area
f(t)	input signal derived from stapes motion (a prescribed harmonic oscillation)
u(x,t)	displacement of the BM averaged over the membrane width

Equation (5) corresponds to a zero trans-BM pressure at the apical end of the cochlea. This condition is preferred to having the x-derivative of the pressure be zero at x = L, because the latter would make the matrix A (see below) singular and hence would require modification of the numercial scheme. For the frequency range of interest (middle to high frequencies) the two conditions are physically equivalent.

A discrete version of the problem is obtained by rendering the equations discrete in x, which yields an initial value problem with t as independent variable. We prefer this to solving a boundary value problem in x, which would result if the order of discretization were reversed.

In addition, starting with the discretization in time would require a priori specification of the time T at which the interval [0,∞) is truncated. If the thus calculated response deviates too much from its stationary value, the computations would have to be repeated with a larger value of T. Our procedure is more flexible: we select a value of T, calculate the response and examine whether a stationary value has been reached. If this is not the case, we increase T and proceed with the already calculated result as new initial value.

The finite element method is most suited to accomplish the discretization in space because of the complexity of the boundary conditions (4) and (5).

We divide the BM length into n intervals (x_{i-1}, x_i), $i = 1,2,\ldots\ldots.n$, with $x_0 = 0$, $x_n = L$. Using Galerkin's principle we reduce Eq. (1) to a system of ordinary differential equations in t:

$$A\,\ddot{\underline{u}} + R\,\dot{\underline{u}} + S\,\underline{u} = \underline{k}, \tag{6}$$

Here, $\underline{u} = (u_1, u_2, \ldots, u_{n-1})^T$ with $u_i = u(x_i,t)$ and $\underline{k} = (-f(t),0,0,\ldots,0)^T$. The matrix A follows from $A = M + C$ with

$$\tag{6a}$$

and

$$C = \frac{1}{2}\,\text{Diag}\,(c_0\Delta_1,\ c_1\Delta_1 + c_1\Delta_2,\ \ldots,\ c_{n-1}\Delta_{n-1} + c_{n-1}\Delta_n), \tag{6b}$$

where m_i stands short for $m(x_i)$, Δ_i for $x_i - x_{i-1}$ and c_i for $2\rho\beta(x_i)/a(x_i)$. The matrices R and S, finally, are obtained by replacing m_i in Eq. (6a) with $r_i = r(x_i,t)$ and $s_i = s(x_i)$, respectively.

Equation (6) will be used to determine u in the gridpoints $x_1, x_2, \ldots, x_{n-1}$. For the endpoint x_n the discretization with respect to x gives

$$\ddot{u}_n = \ddot{u}(x_n,t) = 0, \tag{7}$$

and thus, with the aid of conditions (2) and (3):

$$u(x_n,t) = 0. \tag{8}$$

We rewrite Eq. (6) as a system of first order differential equations by introducing the auxiliary variable $\underline{v} = \dot{\underline{u}}$. In block matrix form, the system reads

$$\begin{vmatrix} I & O \\ O & A \end{vmatrix} \begin{vmatrix} \dot{\underline{u}} \\ \dot{\underline{v}} \end{vmatrix} = \begin{vmatrix} O & I \\ -S & -R \end{vmatrix} \begin{vmatrix} \underline{u} \\ \underline{v} \end{vmatrix} + \begin{vmatrix} O \\ \underline{k} \end{vmatrix} \tag{9}$$

The initial condition is found from Eqs. (2) and (3):

$$\begin{bmatrix} \underline{u} \\ \underline{v} \end{bmatrix} = \begin{bmatrix} \underline{0} \\ \underline{0} \end{bmatrix} \text{ at } t = 0 . \tag{10}$$

We solve u and v from the initial value problem (9), (10) using an explicit fourth order Runge-Kutta scheme with a variable-step routine. This method provides a good compromise as regards computational speed and stability properties. The quantity used for a comparison with experimental data and with other calculations is the frequency spectrum of the BM velocity, which is computed from $v(x,t)$ by means of a Fast Fourier Transform technique.

3. NUMERICAL RESULTS

We have applied the method outlined in the previous section to four model examples.

- Linear, passive [$r(x,t) = r_0(x) > 0$]. This is a testcase which may point out errors in the numerical method or in the implementation. The results of our approach have been compared with those of a finite element solution method in the frequency domain (Borsboom, 1979) for the standard 1D cochlea model. The two responses are in excellent agreement once the transients in the time domain response have become sufficiently small; this takes about 1-2 sec (cochlea time).

- Nonlinear, passive. Following Hall (1974), we set $r(x,t) = r_0(x) \{ 1 + 0.1 \ v^2(x,t) \}$. This type of (compressive) nonlinearity does not give any problems in the computations. Our results compare favourably with those of Hall.

- Linear, active. The example chosen is Kim et al.'s (1980) model, in which the resistance is negative on a part of the BM length. We did not succeed in solving this locally active model. The response becomes unstable before a harmonic response has been reached. The reason is that the transients caused by the onset of the input signal at $t = 0$ are amplified in the region of negative BM resistance. The response enters the region of instability of the numerical method and 'explodes'. Since the proposed Runge-Kutta scheme is very robust, it seems not feasible to solve this type of model in the time domain.

- Nonlinear, active. To our best knowledge, the only published model example in this class is the coupled chain of Van der Pol oscillators (Van Netten, 1982; Van Netten and Duifhuis, 1983). The resistance has the form $r(x,t) = -r_0(x) + r_1(x) \ v^2(x,t)$, with $r_0(x)$ and $r_1(x)$ positive. Fig. 1 shows results of the present approach for this model. The BM response is very much different from the one calculated by Van Netten and Duifhuis,

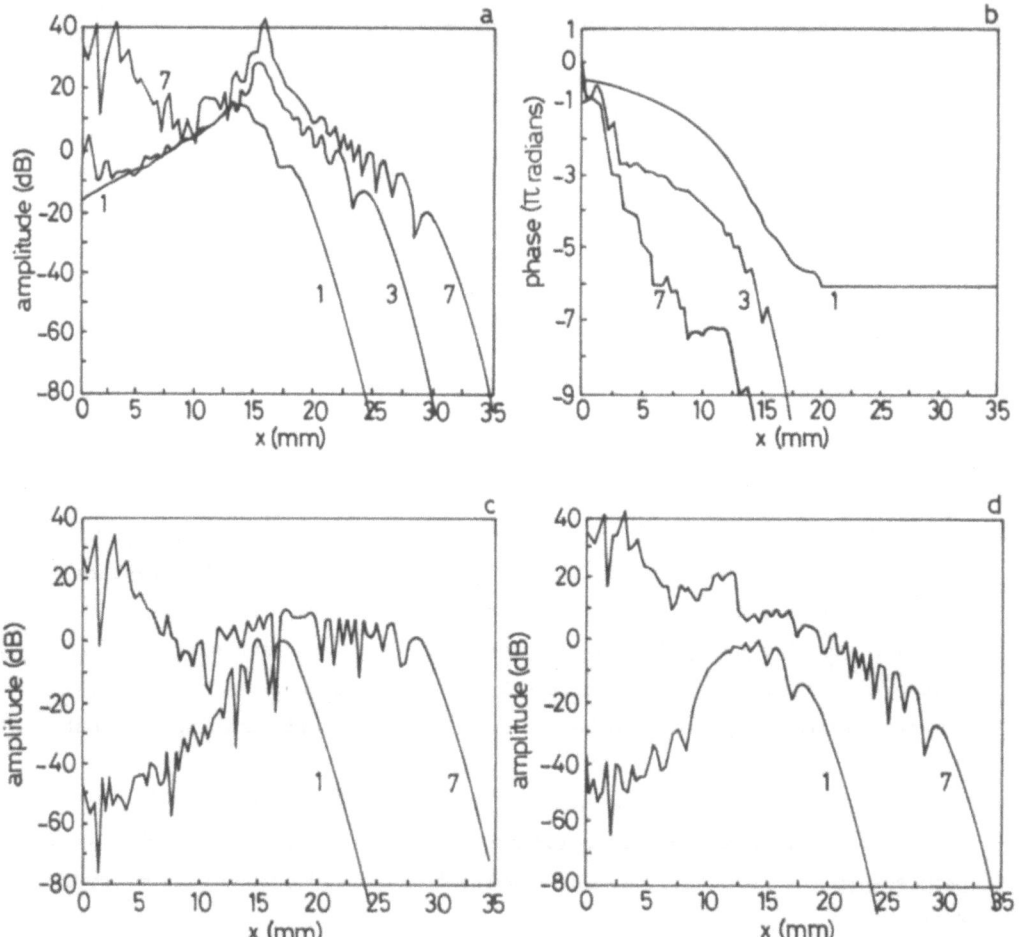

Fig. 1. Response of Van Netten and Duifhuis'(1983) nonlinear active model,
calculated by the approach of section 2.
Input: harmonic stapes oscillation (frequency 2 kHz, amplitude 10^{-5} mm/ms).
Output: BM velocity in the frequency domain, normalized to the amplitude of the
stapes velocity.

a: amplitude of first harmonic (2 kHz), after 1, 3, 7 ms.
b: phase of first harmonic (2 kHz), after 1, 3, 7 ms.
c: amplitude of dc component, after 1 and 7 ms.
d: amplitude of second harmonic, after 1 and 7 ms.
We used 100 equidistant grid points on the BM, and 512 samples per period of
oscillation in our FFT routine.

Further parameters (after Van Netten, 1982): $L = 35$ mm, $\beta = 1$ mm, $a = 1$ mm^2,
$\rho = 1$ mg/mm^3, $m = 0.5$ mg/mm^2, $r = \{-3.5 + 10,000\ v^2\}\ exp(-0.15x)$ mg/mm^{2}ms,
$s = 10,000\ exp(-0.3x)$ mg/mm^{2}ms^2, with x in mm and v (BM velocity) in mm/ms.

especially in the region before the amplitude peak has been reached. Our response saturates in this region owing to the combined effect of activity and nonlinearity.

We suggest the following physical explanation for our results. The negative resistance of the BM produces energy. This creates continuously two series of waves, one going to the apex, the other to the base of the cochlea. Both types of waves are amplified in the stiffness-controlled (basal) part of the BM, which implies that the BM amplitude is increased at every point, but most markedly in the basal region. The compressive nonlinearity ensures that the response remains bounded.

We are quite sure that our calculations are correct, since the results were insensitive to changes in discretization both in space and time (We found no appreciable difference with the approach of Fig. 1 if we doubled the number of grid points on the BM and halved the accuracy of the Runge-Kutta variable-step method). The ripples in the curves are due to the non-local nature of the activity; they disappear if only a small portion of the BM, say just before the amplitude peak, has a negative resistance.

Van Netten and Duifhuis obtained their response directly in the frequency domain by linearizing the time-domain equations with the technique of harmonic balance. This asymptotic technique, which ignores non-harmonic components of the frequency spectrum, is apparently not valid for the model at issue.

4. CONCLUSIONS

We have developed a numerical method to solve 1D cochlea models in the time domain. The method adequately determines the response of passive models, whether linear or nonlinear. It is not suited for linear active models, however, owing to enhancement of transients before a stationary response has been reached. For this type of model we advocate a (non-asymptotic) solution method in the frequency domain. The most interesting case, models having both nonlinear and active BM properties, seems to be handled quite well by the present method.

The intent of this work was to provide a frame of reference for asymptotic solution methods. The developed method indeed appears to be able to decide about the validity of asymptotic approaches applied to cochlea models. As for the main model example treated (Van Netten and Duifhuis, 1983) we found that the harmonic balance technique which linearizes the model equations is inappropriate because it suppresses reflection phenomena. Moreover, our results

160

show that the discussed representation of nonlinear and active properties by
a series of coupled Van der Pol oscillators yields a response that is not in
conformity with experimental data.

Acknowledgements

*We are indebted to Prof. E. de Boer (Amsterdam) and Ir. A. Segal (Delft) for
valuable discussions and comments.*

REFERENCES

Boer, E. de (1983a). No sharpening? A challenge for cochlear mechanics,
    J. Acoust. Soc. Am. 73, 567-573.
Boer, E. de (1983b). On active and passive cochlear models - Towards a gener-
    alized analysis. *J. Acoust. Soc. Am.* 73, 574-576.
Borsboom, M.J.A. (1979). *Linear and nonlinear one-dimensional cochlear models,*
    Eng. D. thesis, Dept. of Mathematics and Informatics, Delft University of
    Technology.
Hall, J.L. (1974). Two-tone distortion products in a nonlinear model of the
    basilar membrane, *J. Acoust. Soc. Am.* 56, 1818-1828.
Kemp, D.T. (1978). Stimulated acoustic emissions from within the human audi-
    tory system, *J. Acoust. Soc. Am.* 64, 1386-1391.
Khanna, S.M. and Leonard, D.G.B. (1982). Basilar membrane tuning in the cat
    cochlea, *Science* 215, 305-306.
Kim, D.O., Neely, S.T., Molnar, C.E., and Matthews, J.W. (1980). An active
    cochlear model with negative damping in the partition: Comparison with
    Rhode's ante- and postmortem observations. In: *Psychophysical, physiolo-
    gical and behavioural studies in hearing.* G. van den Brink and F.A.
    Bilsen (eds.), Delft Univ. Press, 7-15.
Neely, S.T. (1981). *Fourth order partition mechanics for a two-dimensional
    cochlear model.* Doctoral dissertation, Washington Univ., St. Louis.
Neely, S.T. and Kim, D.O. (1983). An active cochlear model showing sharp
    tuning and high sensitivity. *Hearing Res.* 9, 123-130.
Netten, S.M. van (1982). *Een actief, niet-lineair cochlea model (in Dutch),*
    M. Sc. thesis, Dept. of Biophysics, Univ. of Groningen.
Netten, S.M. van and Duifhuis, H. (1983). Modelling an active nonlinear
    cochlea. In: *Proc. Mechanics of Hearing,* E. de Boer and M.A. Viergever
    (eds.), Delft University Press/Martinus Nijhoff Publishers, pp. 143-151.
Rhode, W.S. (1971). Observations of the vibration of the basilar membrane in
    squirrel monkeys using the Mössbauer technique, *J. Acoust. Soc. Am.* 49,
    1218-1231.
Rhode, W.S. (1978). Some observations on cochlear mechanics, *J. Acoust. Soc.
    Am.* 64, 158-176.
Sellick, P.M., Patuzzi, R., and Johnstone, B.M. (1982). Measurement of basilar
    membrane motion in guinea pig using the Mössbauer technique, *J. Acoust.
    Soc. Am.* 72, 131-141.
Viergever, M.A. (1980). *Mechanics of the inner ear - a mathematical approach.*
    Delft University Press.
Viergever, M.A. and Diependaal, R.J. (1983). Simultaneous amplitude and phase
    match of cochlear model calculations and basilar membrane vibration data.
    In: *Proc. Mechanics of Hearing,* E. de Boer and M.A. Viergever (eds.),
    Delft University Press/Martinus Nijhoff Publishers, pp. 53-61.

Section V

Nonlinear micromechanics

EXPERIMENTAL AND TOPOGRAPHIC MORPHOLOGY IN COCHLEAR MECHANICS

Luboš Voldřich

Institute of Experimental Medicine
Czechoslovak Academy of Sciences

ABSTRACT

*The vital basilar membrane is anisotropic, that is why its movements differ
from the postmortally changed one. The spiral ligament is, due to its elasticity, directly involved in cochlear mechanics. By modelling the movement of
the cochlear partition segment, the lever-mediated transformation from BM displacement to the displacement of hair cell stereocilia is demonstrated. These
findings have been documented microcinematographically.
In a three-dimensional model, topographic relations of the supporting system
showing the interconnection of individual radial segments are presented.
During basilar membrane motion a complex pattern arises in the reticular lamina.
This augments the contrast in displacement of neighbouring hair cells.*

1. VITAL BASILAR MEMBRANE RESPONSE

To model the processes of cochlear mechanics we proceed from the principles of

acoustics and from histological information on the general structure of the

cochlear organ and on the mechanical characteristics of its particular struc-

tures. A century of research into the organ of hearing has yielded a wealth of

information. Much of it has passed unnoticed, some of it has tended to be em-

phasized or even overrated. Often, in studying the physiological processes of

the cochlea, findings made on postmortem-altered tissues are used, e.g. in

modelling basilar membrane (BM) motion. A fresh BM responds to pressure on a

small part of its surface in a way quite different from that of a chemically

fixed or postmortem-altered membrane. A living membrane exhibits marked aniso-

tropy, a dead one soon loses its mechanical orientation. The high sensitivity

of BM mechanical properties to disorders in intracochlear homeostasis was con-

firmed by Khanna and Leonard (1982). The BM radial fibres keep the shape of

deformation of the living membrane confined to a narrow transverse groove

whereas a fixed membrane or one examined a few dozen minutes after death be-

comes deformed over a wide crater-shaped circular area as described previously

(Voldřich, 1978). Structurally speaking, anisotropy is conditional upon the

combined action of the elastic basic substance and the radial bundles of

fibres which are elastic in bending but noted for tensile strength. The BM

anisotropy is conducive not only to acute tuning and, consequently, to a high-

degree frequency analysis already in the cochlea, but also to greater sensi-

tivity to threshold intensities as less fluid displacement by the stapes is

needed for deviation.

2. ROLE OF THE SPIRAL LIGAMENT IN COCHLEAR MECHANICS

The tensile strength of the radial fibres will not permit their prolongation, which is why the BM width remains unchanged in deviation. In instrumental manipulation, deviating the BM, we are able to see a live cochlear partition constituting a functional entity together with the spiral ligament. The BM can deviate thanks to the elasticity of the ligament which acts as a yielding and elastic structure rather than as a rigid support for the membrane periphery (Voldřich, Úlehlová, 1982).

Comparative morphology of the cochlea in different mammals including man showed the BM and the spiral ligament as being distinctly related to the function of hearing provided they are studied as a whole. The essential point there is not the absolute width of the BM but rather the ratio of its width with regard to that of the spiral ligament (on the radial plane). This ratio shows a significant correlation with the cochlear tonotopy and with the range of hearing in different species (Burda, 1983).

The elasticity of the spiral ligament should be taken into account in modelling cochlear mechanics. The BM is to be seen as a membrane stretched between the bony lamina and the ligament. Only so we can obtain a model of the conditions proper to a living mammalian organ. The mode of deviation in a fresh as distinct from a postmortem-altered cochlear partition, as well as the role the spiral ligament has to play in cochlear mechanics have been documented microcinematographically (Voldřich, Úlehlová, 1981).

3. TRANSFER OF BASILAR MEMBRANE MOTION TO STEREOCILIA

Once we accept the idea of a mechanical frequency BM response along narrow radial segments, we can easily imagine the whole of the organ of Corti responding as a chain of mutually independent transverse radial segments. One such segment would be made up of one inner and one outer pillar and three supporting Deiters' cells, and of one inner and three outer hair cells. The fulcrum would be at the point of BM insertion to the bony lamina of the modiolus. The motion of the BM causes a shift of the reticular lamina against the tectorial membrane, the fulcrum of which is the edge of the spiral limbus.

Let us suppose that the complex of supporting cells and reticular lamina is a rigid structure, similarly as the BM pars tecta; together with the pillars they form a fixed triangle. Let us also regard the tectorial membrane as rigid. By creating a model of relationship between the hypothetical segment of the organ of Corti and the tectorial membrane we can conceive an idea of how the BM deviation is transmitted to that of the stereocilia. It appears then that the angular displacement of the stereocilia is directly proportional to the BM displacement and to the height of the organ of Corti, and indirectly proportional to the length of the stereocilia and to the width of the displaced part of the BM. If we apply those relationships to the actual morphometric readings we can see that this lever-mediated transformation augments the BM displacement several times resulting in a relatively large displacement of the stereocilia. For instance, in the second coil of the guinea-pig cochlea the angular displacement build-up is about seven-fold.

However, the actual topographic relationships of the cells of the organ of Corti are too complex to warrant any oversimplified ideas of independent radial segments. For orientation's sake let us choose the direction of the radial fibres of the BM. Scrutinizing the surface of the reticular lamina we can see no distinct structure in this direction. What we can see in the cellular pattern, though, is that the triad of outer hair cells is turning away from the radial direction towards the cochlear apex exactly as the tectorial membrane fibres do. The cuticular plates of the hair cells are turned proportionally, thus preserving the orientation of the stereocilia with regard to that of the tectorial membrane fibres. The hypothetical radial segment is, therefore, spatially curved, its base being radial and straight, its top in the reticular lamina inclined, with the peripheral portion turned towards the apex of the cochlea.

4. NON-LINEARITY IN COCHLEAR MICROMECHANICS

Unfortunately, the hypothesis of independent segments is thwarted by other topographic relationships, too. We can hardly disregard the fact, obvious enough from Retzius' drawings of the organ of Corti as well as from present-day pictures produced by scanning electron microscopy, that the base of each pair of pillar cells is not actually situated on the same radial axis of the BM, but that the inner pillars are shifted towards the cochlear base. The supporting Deiters' cells send out phalangeal processes to the reticular lamina two to three outer hair cells apically, whereas the hair cells are turned away in the basal direction (Fig. 1).
The phalangeal axes and those of the hair cells subtend an angle of up to 60°. Thus, one or even two outer hair cell cuticular plates are situated between the hair cell cuticular plate and the phalangeal plate arising from the same supporting cell. Beside obliquity in the base-apex sense, there is also a different inclination in the modiolar plane, i.e. from the centre to the periphery of the cochlea.

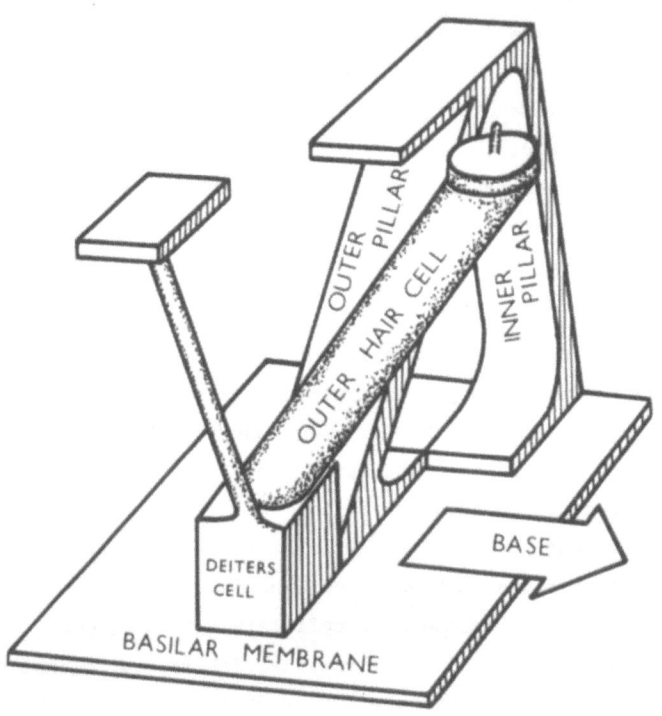

Fig. 1. Drawing of the three-dimensional topographic relationship of the sup-porting and outer hair cells in the Corti organ.

Thus, if we construct a rigid frame model with mutually interconnected elements, displacement at a given point on the BM will set the whole system in motion, in which case the purpose of acute BM tuning is lost.

We have constructed a model made of rigid elements representing the frame of the organ of Corti and loosely hinged to the BM and the reticular lamina rather like articulated joints. The BM displacement is transmitted to the reticular lamina in a complex sort of pattern. By altering the degree of rigidity or freedom in the articulations, the BM displacement is transmitted to the reticu-lar lamina in different modes. Different parts of the lamina are displaced differently. A BM lift transmitted by way of the outer pillar will cause the lamina to shift mainly towards the modiolus, while a phalangeally mediated lift of the BM will cause the adjacent sector of the lamina to move up. This, in turn, affects the neighbouring sectors on the reticular lamina, thus causing different displacements of the stereocilia. It follows from the function of the above described model that on the reticular lamina there is a growing contrast of displacements of hairs of the neighbouring sensory cells. This might be considered to be a mechanism for a more precise and more acute localization of stimulation of auditory cells.

REFERENCES

Burda, H. (1983). Relation of the spiral ligament form to the auditory function. In: *Kochleafunktion*, edited by H. Jakobi, K.D. Kühl, and P. Lotz (Barth, Leipzig), in press.
Khanna, S.M. and Leonard, D.G.B. (1982). Basilar membrane tuning in the cat cochlea, *Science* 215, 305-306.
Voldřich, L. (1978). Mechanical properties of basilar membrane, *Acta Otolaryngol.* 86, 331-335.
Voldřich, L. and Úlehlová, L. (1981). *Cochlear Mechanics*. Colour film, 17 min. made by Krátký film - Praha.
Voldřich, L. and Úlehlová, L. (1982). The role of the spiral ligament in cochlear mechanics, *Acta Otolaryngol.* 93, 169-173.

RESULTS FROM A COCHLEAR MODEL UTILIZING LONGITUDINAL COUPLING

Y.C. Jau, C.D. Geisler

University of Wisconsin-Madison
USA

ABSTRACT

A one-dimensional computer simulation of the cat cochlea was developed utilizing a one-degree-of-freedom model of the cochlea partition, with the magnitude of the damping between the reticular lamina and the tectorial membrane dependent upon the envelope *of basilar-membrane displacement integrated over a* longitudinal *region. Aspects of the model's output mimic auditory-nerve fiber activity. The slopes of stimulus-response curves for single-tone stimuli decrease as frequency is increased above characteristic frequency. Appropriate two-tone suppression is observed when the suppressing tone's frequency is either above or below that of an excitatory tone at characteristic frequency. Nonlinear longitudinal coupling in the cochlea is suggested.*

1. INTRODUCTION

Models of the cochlea typically have assumed that the basilar partition can be divided up into discrete segments that have no coupling between them other than through the fluid in the scalae. Yet evidence has been accumulating that there is appreciable longitudinal coupling among elements of the organ of Corti. Many years ago, von Békésy (1960) reported that the tectorial membrane had considerable stiffness in the longitudinal direction, pivoting about the osseous spiral lamina like a book cover. More recently, Javel, Geisler and Ravindran (1978) found that the strength of the suppression exerted on a primary auditory fiber's rate of response to an excitatory tone by a higher-frequency suppressive tone dropped off exponentially with inferred spatial separation between the respective characteristic places. Another set of experiments on auditory-nerve fibers (Robertson and Johnstone, 1981) has indicated that exposing the ear to a high-intensity tone whose frequency is above best frequency can cause a reduction in two-tone suppression without changing the response pattern to a single tone. The most likely explanation for this reduction is that a longitudinal electromechanical effect normally caused by the suppressive tone is fatigued or weakened by the high-intensity tone.

From this and other evidence, we are convinced that realistic models of the cochlear partition must include longitudinal coupling. This paper describes the results obtained with a model containing such coupling. Our specific hypothesis is that the presentation of sound causes the tectorial membrane to undergo static deformations toward the reticular lamina. This static deformation decreases the thickness of the subtectorial space, thereby increasing the damping resistance of the fluid confined to that space. This nonlinear effect is assumed to be proportional to the envelope of the displacement and to extend longitudinally along the cochlea with exponentially decreasing strength.

2. THE MODEL

Our model is a one-dimensional computer simulation of a 22.5-mm cat cochlea containing 100 sections (Jau, 1983). For each section, the scalar fluids are represented by a mass and the cochlear partition by the one-degree-of-freedom model shown in Fig. 1. This model, derived from Allen (1980), represents the basilar membrane with a spring-mass (k_b – m_b) system which is coupled to a rigid rotatable tectorial membrane by viscous fluid (r_p) and hair-cell cilia stiffness (k_c). The radial shear lever gain (g) relates vertical basilar-membrane displacement (ξ_b) with its effective radial displacement.

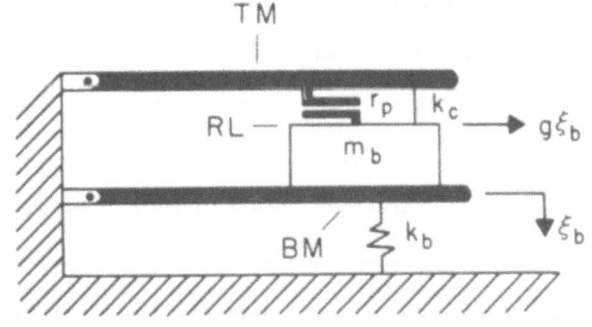

Fig. 1. Model of cochlear partition used. TM--tectorial membrane, BM--basilar membrane, RL--reticular lamina. Other symbols defined in text. (Adapted from Allen, 1980.)

Due to our assumption that static deflections of the tectorial membrane occur
and extend for appreciable longitudinal distances, we represented the subtectorial fluid as a longitudinally controlled nonlinear damper. Specifically, we assumed that

$$r_p(x,t) = r_{po}(1 + \beta f(x,t))$$
(1)

where r_{po} is a rest damping, β is a weighting factor, and

$$f(x,t) = \int_{x-\alpha}^{x+\alpha} D(y,t) \exp(-|y - x|/\lambda) \, dy$$
(2)

where $D(y,t)$ is the envelope of basilar-membrane displacement at a particular
place and time, λ is the longitudinal space constant, and α is the limit of
longitudinal coupling.

All simulations were for 12 msec of simulated time. Values of model parameters are given in Jau (1983).

3. RESULTS

The envelopes of basilar-membrane displacement in response to several intensities of a 2.5-kHz tone are shown in Fig. 2a. As the intensity of the
stimulus increases, the degree of peakiness decreases and the location of
the maximum response (the "characteristic place") shifts toward the base.
For the formulation chosen, the 40-dB response has about the maximum amount
of peakiness that we could achieve in 12 msec of simulated time.

The amplitudes of the basilar-membrane's vibrations at the 16.2-mm point in
response to several different frequencies and amplitudes are shown in Fig. 2b.
At the characteristic frequency, 2.5 kHz, the amplitude curve has a slope of
less than unity between 40 and 60 dB and shows even a smaller slope for higher
intensities. At lower frequencies, the slopes are greater, with the low-

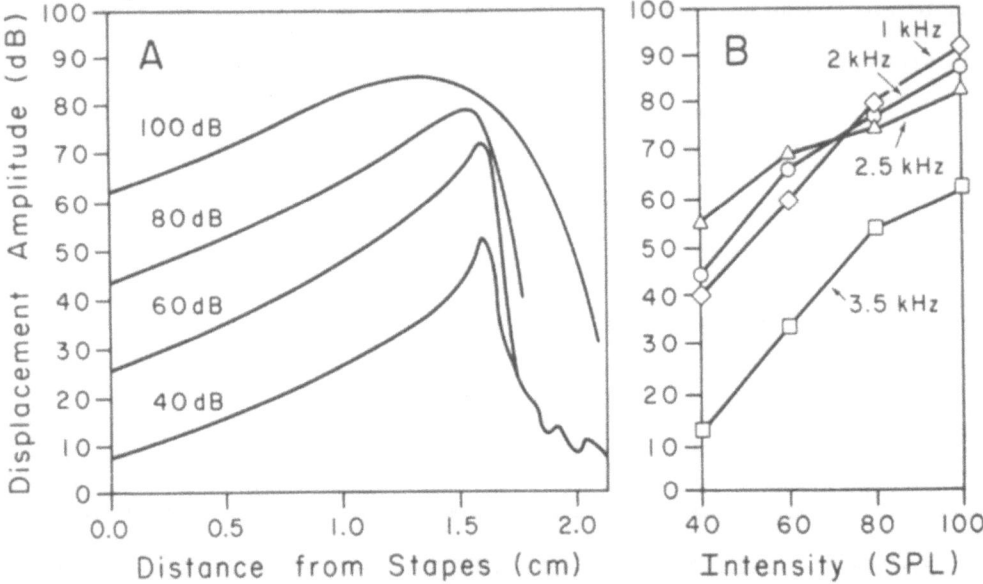

Fig. 2. Basilar-membrane displacement (arbitrary scale). Part (a): vs. cochlear distance for 2.5-kHz tone stimulation at several intensities; part (b) vs. stimulus intensity for several frequencies, at x = 16.2 mm (CF = 2.5 kHz). β = 300, α = 4.5 mm, λ = 1 mm (a) or 10 mm (b).

intensity slopes approaching unity. The low-intensity slope for the 3.5-kHz curve is also close to unity, but the amplitude of response is much lower. Therefore, at 80 dB SPL, a stimulus intensity that is great enough to produce response amplitudes in the range of 50 dB, the higher-frequency curve has a slope that is even less than that of the characteristic frequency in that amplitude range. The cause of this latter slope reduction is the spread of damping that occurs from more basal points of the cochlea, where relatively large displacement amplitudes occur in response to the high-intensity signal. Even though large response amplitudes are achieved at this cochlear place (x = 16.2 mm) with the lower-frequency signals, the fact that it is well basal to the characteristic places for these frequencies means that its corresponding response amplitudes are controlled by the membrane stiffness, and not by the damping (Hubbard and Geisler, 1972), and hence behave nearly linearly.

The type of behavior shown in Fig. 2b is very reminiscent of auditory-nerve fiber activity, for which different rate-intensity curves usually have different slopes depending on the stimulus frequency. At frequencies well below characteristic frequency, these slopes have a common value, while for stimulus frequencies near and above characteristic frequency, the values of the slopes decrease monotonically with frequency (Geisler, Rhode and Kennedy, 1974; Sachs and Abbas, 1974).

The response of the model to the simultaneous presentation of two tones is shown in Fig. 3a. In this case, the amplitude of the higher-frequency signal (f2 = 2.5 kHz) is 28 dB higher than that of the lower-frequency tone (f1 = 1.6 kHz). For comparison purposes, the response to the 1.6-kHz tone presented by itself is also shown. Notice that the addition of f2 to f1 causes a reduction of about 4 dB in the maximum value of the membrane's response at the characteristic place for f1 (17.8 mm). Response amplitudes to f1 alone and to

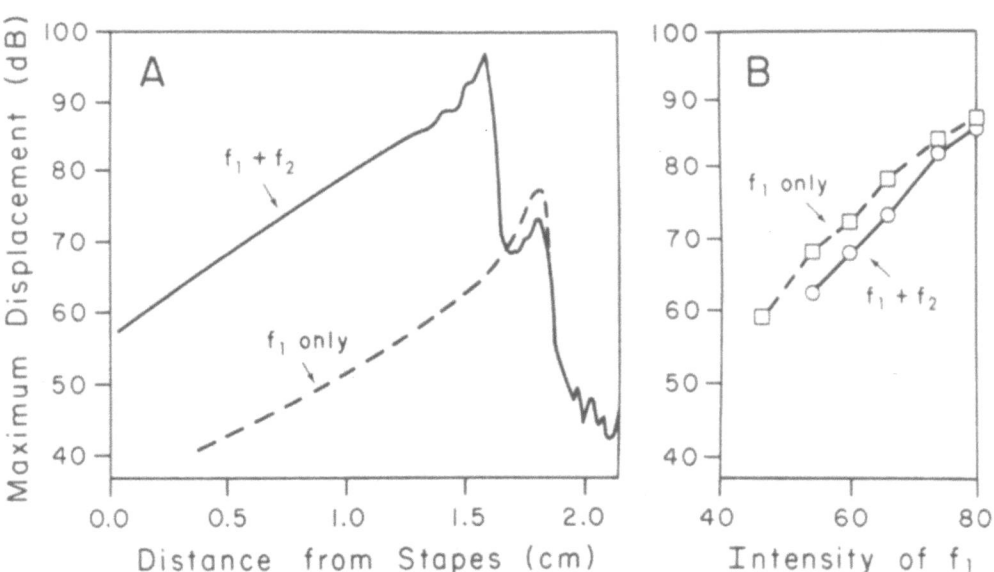

Fig. 3. *Maximum basilar-membrane displacement. Part (a): vs. cochlear distance for two-tone stimulation (f1 = 1.6 kHz @ 66 dB SPL; f2 = 2.5 kHz @ 94 dB SPL); part (b): vs. intensity of f1 (1.6 kHz), presented with and without f2 (2.5 kHz @ 94 dB SPL), at x = 17.8 mm (CF = f1). β = 100, α = 2.25 mm, λ = 1 mm.*

f1 and f2 presented simultaneously are shown in Fig. 3b. Notice that the two curves are approximately parallel, especially over the low-intensity region. As the frequency of f2 increases relative to f1, the amount of suppression decreases (Jau, 1983).

Auditory-nerve fibers also display suppression of response amplitudes when the suppressive tone f2 has a frequency well below that of f1 (Sachs, 1968). A similar effect is observed in the model. The curve in Fig. 4 shows that the

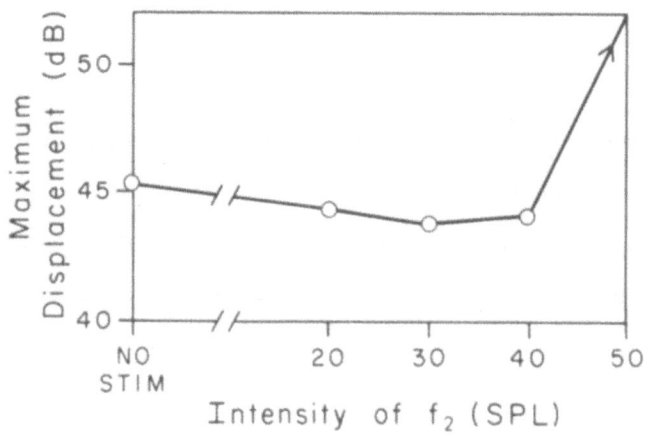

Fig. 4. Maximum basilar-membrane displacement vs. intensity of f2 (1 kHz), presented with f1 (2.5 kHz @ 20 dB), at x = 16.2 mm (CF = f1). β = 1000, α = 4.5 mm, λ = 10 mm.

magnitude of the response to two tones, at the characteristic place for the higher-frequency tone (f1 = 2.5 kHz), shows non-monotonic behavior with increases in the intensity of the low-frequency tone (f2) The response amplitude decreases with increases in the strength of a low-intensity f2, but grows with increases in f2's intensity at higher levels. The response attenuation at the lower intensities is due to the spread of damping power from the apical sections excited by f2, while the increase in the response amplitude at the higher intensities is due to the excitatory effects of the low-frequency tone at the recording point.

4. DISCUSSION

The idea of longitudinal coupling is not new but has been advanced by a
number of other investigators. More than a decade ago, Lynn and Sayers (1970)
proposed that the inner and outer hair cells, innervating different sections
of the cochlea, were coupled neurally. More recently, Zwislocki (1975) as
well as Dolmazon and Boulogne (1982) have proposed that the drive on the
afferent nerve fibers located at a particular spot is the difference between
activity due to the inner hair cells and that due to more basally located
outer hair cells. We believe that this difference hypothesis is unable to
account for such phenomena as two-tone suppression, for the period histograms
of primary-fiber discharges taken under conditions of strong high-frequency f2
suppression show no evidence of synchronizing to the suppressing frequency
(Javel, Geisler and Ravindran, 1978). In a manner somewhat similar to ours,
Kletsky and Zwislocki (1979) included longitudinal coupling in the tectorial-
membrane portion of an electronic model of the cochlea, but the model that
they used was linear. More closely related in spirit to our model is a reso-
nating-reed model of the cochlea developed by these same authors (Zwislocki
and Kletsky, 1980), in which the introduction of nonlinear coupling by means
of a longitudinally stretched rubber band caused two-tone suppression.

The results presented here show that the introduction of displacement-depen-
dent longitudinal coupling into a cochlear model causes several nonlinear
effects that are also observed in the discharge patterns of auditory-nerve
fibers. For one, amplitude-intensity curves for single-tone stimuli take on
slopes of decreasing value as frequency is increased above characteristic
frequency. Secondly, two-tone suppression of the appropriate character is
observed when the frequency of the suppressing tone is either above or below
that of an excitatory tone at characteristic frequency. Quantitative comparisons between the model data and the neural data were not made, because our

oversimplified model's responses to tonal stimuli are not nearly peaked enough at low intensities to accurately represent basilar-membrane vibrations. Nevertheless, the trends in the model's results are unequivocal and clearly demonstrate the surprising range of realistic effects produced by nonlinear longitudinal coupling. Similar mechanisms may be at work in the cochlea.

Acknowledgements

This research was supported by NIH Program Project Grant NS-12732 and by the University's Computer-Aided Engineering Center.

REFERENCES

Allen, J.B. (1980). Cochlear micromechanics—A physical model of transduction, *J. Acoust. Soc. Am.* 68, 1660-1670.

Békésy, G. von (1960). *Experiments in Hearing* (McGraw-Hill, New York), p. 468.

Dolmazon, J.M., and Boulogne, M. (1982). Interaction phenomena in a model of mechanical to neural transduction in the ear, *Speech Comm.* 1, 55-73.

Geisler, C.D., Rhode, W.S., and Kennedy, D.T. (1974). Responses to tonal stimuli of single auditory nerve fibers and their relationship to basilar membrane motion in the squirrel monkey, *J. Neurophysiol.* 37, 1156-1172.

Hubbard, A.E., and Geisler, C.D. (1972). A hybrid-computer model of the cochlear partition, *J. Acoust. Soc. Am.* 51, 1895-1903.

Jau, Y.C. (1983). Nonlinearities in vibration pattern of a cochlear partition model. M.S. Thesis, Univ. of Wisconsin-Madison, Madison, Wis.

Javel, E., Geisler, C.D., and Ravindran, A. (1978). Two-tone suppression in auditory nerve of the cat: Rate-intensity and temporal analyses, *J. Acoust. Soc. Am.* 63, 1093-1104.

Kletsky, E.J., and Zwislocki, J.J. (1979). Cochlear-microphonic versus hair-cell tuning in the cochlea, *J. Acoust. Soc. Am.* 65, S84.

Lynn, P.A., and Sayers, B. McA. (1970). Cochlear innervation, signal processing, and their relation to auditory time-intensity effects, *J. Acoust. Soc. Am.* 47, 525-533.

Robertson, D., and Johnstone, B.M. (1981). Primary auditory neurons: Nonlinear responses altered without changes in sharp tuning, *J. Acoust. Soc. Am.* 69, 1096-1098.

Sachs, M.B. (1968). Stimulus-response relation for auditory-nerve fibers: Two-tone stimuli, *J. Acoust. Soc. Am.* 45, 1025-1036.

Sachs, M.B., and Abbas, P.J. (1974). Rate versus level functions for auditory-nerve fibers in cats: Tone-burst stimuli, *J. Acoust. Soc. Am.* 56, 1835-1847.

Zwislocki, J.J. (1975). Phase opposition between inner and outer hair cells and auditory sound analysis, *Audiol.* 14, 443-455.

Zwislocki, J.J., and Kletsky, E.J. (1980). Micromechanics in the theory of cochlear mechanics, *Hearing Res.* 2, 505-512.

AN INTERPRETATION OF THE SHARP TUNING OF THE BASILAR MEMBRANE MECHANICAL RESPONSE

S.M. Khanna, D.G.B. Leonard

Columbia University
New York, USA

ABSTRACT

It has been shown (Khanna and Leonard, 1982) that sharply tuned basilar membrane (BM) responses are seen when the trauma produced in opening the cochlea is reduced to a minimum. In these animals, ear canal sound pressure levels (SPL) show a pressure minimum of 2 to 20 dB in the frequency region 10-30 kHz. The purpose of this paper is to show that the sharp BM tuning is independent of SPL minimums. An interpretation of the sharply tuned mechanical response is given. Its origin is in the sharply tuned mechanical response of the stereocilia bundles of the hair cells.

1. INTRODUCTION

In the frequency region above 10 kHz most of the incident acoustic waves in the ear canal are reflected by the tympanic membrane (TM) at the far end. The interference between the incident and reflected waves produces standing waves (Stinson and Shaw, 1982; Khanna, 1982). A pressure minimum can occur only at that distance from the point of reflection where the phase angle between the two waves is an integral multiple of 2π radians. It is therefore possible for a minimum to occur at the probe microphone position (several mm from the TM) but not at the TM where the reflection occurs. As a consequence, if BM vibrations are referred to sound pressure measured by the probe microphone, the pressure minimum may enhance the real sharpness of the BM tuning if both occurred at the same frequency.

An obvious way to bypass the problem of sound pressure measurement is to refer the basilar membrane (BM) vibration amplitude to the electrical input of the acoustic transducer. This can be done in our experiment since the Sokolich acoustic driver frequency response is flat throughout the frequency region of interest (1 to 30 kHz). It has been clear to us since 1980 (Khanna and Leonard, 1981B) that sharp tuning of the BM can be demonstrated under conditions in which there are no standing wave artifacts.

2. RESULTS

In a few experiments the relative SPL measured by the probe microphone (constant driver voltage) in the 10-30 kHz region was relatively flat ($\pm$ 5 dB) and it did not vary in a way which would effect the sharpness of the BM tuning

(Fig. 1A). Electrical input applied to the acoustic transducer to produce a basilar membrane vibration of 10^{-8} cm is shown (Fig. 1B). A sharply tuned BM response with a peak-to-tail (1 kHz) ratio of 30 dB is seen. Two sets of responses measured roughly an hour apart are shown (Cat 9/24/80).

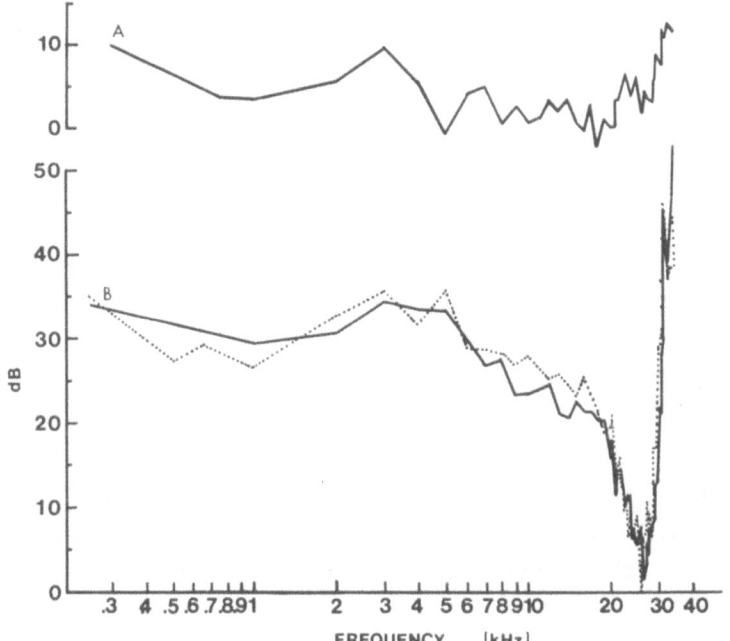

Figure 1.

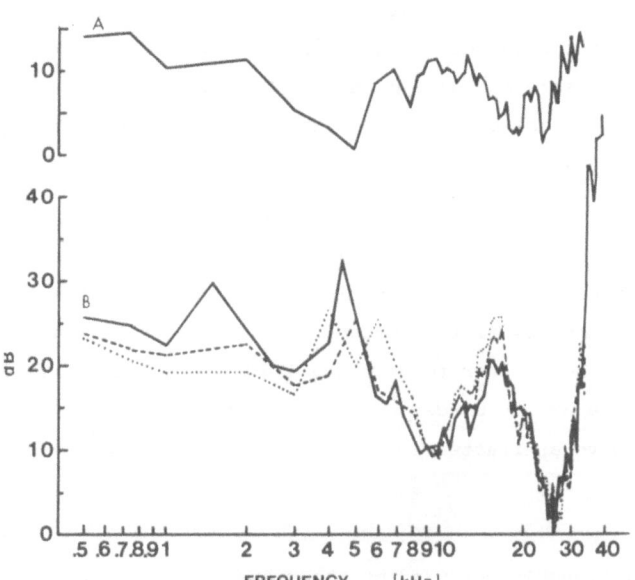

Figure 2.

In nine out of twelve experiments only shallow (5-7 dB) broad minimums similar to the one shown in Fig. 2A were observed in the calibration test.

The contribution of these in sharpening the observed BM response would be minimal. Relative magnitude of the electrical input to the transducer required to
produce a basilar membrane response of 10^{-8} cm is shown in Fig. 2B. A sharp
tuning of the response with a CF of 26 kHz and a peak-to-tail ratio of 20 dB is
seen (Cat 3/10/81).

3. REVIEW OF TUNING PROPERTIES

As stated earlier (Khanna and Leonard, 1982), the sharp tuning seen in our experiments is a direct consequence of reduction of trauma to the cochlea. The
interpretation of the sharply tuned BM response has been discussed in detail
elsewhere (Khanna, 1983A,B), it is summarized below.

When SPL required to produce a constant BM vibration amplitude is plotted as a
function of frequency two types of response components are seen.

- A robust linear response is observed even in the presence of extensive
  trauma which may cause a loss in round window microphonic potential that
  exceeds 50 dB (Khanna and Leonard, 1981A). The SPL required is relatively
  constant up to a cut off frequency beyond which it increases steeply. The
  cut off frequency varies with the position on the cochlea. This response
  has been measured by a large number of investigators and is well understood (for a review see Rhode, 1980; Zwislocki, 1981).

- An extremely fragile nonlinear response which is sharply tuned and located
  just below the cut off frequency of the linear response (Khanna and
  Leonard, 1982; Sellick et al., 1982). It is seen only in cochleas in which
  trauma is minimal.

A comparison of BM mechanical tuning curves, hair cell dc receptor potential
tuning curves (Sellick and Russell, 1978) and frequency tuning curves of auditory nerve fibers (M.C. Liberman, personal communication) all with similar
CF show important similarities and differences (see Khanna, 1983, for details).

- They all have similar shapes and Q_{10} values. The peak height in dB (CF to
  0.5 CF) is much smaller for the BM, 35 dB as compared to 80 dB for the
  nerve fibers and the hair cells.

- Trauma affects all three in the same general way: (i) CF moves to lower
  frequency, (ii) sensitivity decreases, (iii) slopes are reduced, (iv)
  sensitivity reduction is larger in magnitude at the hair cell than at the
  BM.

- All three have nonlinear responses in the sharply tuned portion. The nonlinearity of the BM response disappears at death (Rhode, 1974). The

mechanical nonlinearity of the hair cell stereocilia bundle has been observed by direct stimulation (Strelioff and Flock, 1982).

This comparison indicates that the sharply tuned nonlinear mechanical response
originates at the hair cell.

4. INTERPRETATION

The resonant properties of the stereocilia bundles of individual hair cells
determine the shape of tuning seen in the sharply tuned tip region of the
hair cell and single nerve fiber tuning curves in the auditory system. The
physical dimensions of the stereocilia (diameter, height, number) are organized
along the length of the auditory sensory organ in such a way that the hair
cells which respond to higher frequencies have stiffer ciliary tufts (Lim,
1980; Tilney and Saunders, 1982). The tuning frequency of each hair cell is
determined by the mechanical parameters of its own stereocilia bundle and the
associated tectorial tissue.

Stimulation of the hair cells occurs via the tectorial membrane attachment to
the tallest row of stereocilia. Due to this mechanical coupling, the sharply
tuned response of the hair cell is reflected in the BM response. The mechanical
properties of stereocilia are nonlinear and this nonlinearity is also reflected
in the sharply tuned portion of the BM response. Trauma reduces the stiffness
of the tallest stereocilia and decouples them from the stereocilia bundle
(Tilney et al., 1982). Loss of stiffness lowers the resonant frequency of the
stereocilia bundle and loss of coupling lowers their sensitivity. The loss of
coupling also prevents the sharply tuned nonlinear response of the stereocilia
from appearing at the basilar membrane.

The principle that the tuning in the inner ear is due to mechanical resonant
properties of the stereocilia may be quite general in nature because in a
large number of animal species (amphibians) the auditory organs lack a basilar
membrane (Wever, 1983). The hair cells rest on a stationary supporting structure. These hair cells can only be stimulated through the vibration of their
ciliary tufts. The sharpness of tuning observed in the auditory nerve fibers
in these ears is comparable to that observed in the mammalian ears. The
frequency selectivity in the amphibian ears can only be due to the mechanical
properties of the stereocilia and the attached tectorial tissue (Capranica,
1978). Mechanical tuning properties of free standing stereocilia in alligator
lizard have been demonstrated (Holton, 1983). In view of the above a recon-
sideration of some of the basic principles of the inner ear mechanics is in
order.

REFERENCES

Capranica, R.R. (1978). Auditory processing in anurans. *Fed. Procd.* <u>37</u>, 2324-2328.

Holton, T. (1983). A quantative analysis of hair-bundle and receptor organ motion in the cochlea of the alligator lizard. *Assoc. Res. in Otolaryngol. Mid-winter Research Meeting* (abstract) pp. 98-99.

Khanna, S.M. (1982). Unpublished observations.

Khanna, S.M. and Leonard, D.G.B. (1981A). Basilar membrane response measured in damaged cochleas of cats. In: *Mathematical modeling of the hearing process*, eds. M.H. Holmes & L.A. Rubenfeld, (Springer-Verlag, N.Y.), pp. 70-84.

Khanna, S.M. and Leonard, D.G.B. (1981B). Laser interferometric measurements of basilar membrane vibrations in cats using a round window approach. *J. Acoust. Soc. Amer.* <u>69</u>, S51.

Khanna, S.M. and Leonard, D.G.B. (1982). Basilar membrane tuning in the cat cochlea. *Science*, <u>215</u>, pp. 305-306.

Khanna, S.M. (1983A). Inner ear function based on the mechanical tuning of the hair cell. In: *Recent developments in hearing science*, ed. C. Berlin, (College Hill Press, Calif.), in press.

Khanna, S.M. (1983B). Interpretation of the sharply tuned basilar membrane responses obtained in the cat cochlea. In: *Hearing and other senses: presentation in honour of E.G. Wever*, ed. R.R. Fay, (Amphora Press, Groton, Conn.), in press.

Lim, D. (1980). Cochlea anatomy related to cochlear micromechanics. A review. *J. Acoust. Soc. Amer.*, <u>67</u>, 1686-1695.

Rhode, W.S. (1974). Measurement of vibration of the basilar membrane in the squirrel monkey. *Annals Oto. Rhino. & Laryngol.*, <u>83</u>, 619.

Rhode, W.S. (1980). Cochlear partition vibration- recent views. *J. Acoust. Soc. Amer.* <u>67</u>, 1696-1703.

Sellick, P.M., Patuzzi, R. and Johnstone, B.M. (1982). Measurement of basilar membrane motion in the guinea pig using the Mössbauer technique. *J. Acoust. Soc. Amer.* <u>72</u>, pp. 131-141.

Sellick, P.M. and Russell, I.J. (1978). Intercellular studies of cochlear hair cells: filling the gap between basilar mebrane mechanics and neural excitation. In: *Evoked electrical activity in the auditory nervous system*, eds. R.F. Naunton & C. Fernandez, (Academic Press, N.Y.), pp. 113-139.

Stinson, M.R. and Shaw, E.A.G. (1982). Wave effects and pressure distribution in the ear canal near the tympanic membrane, *J. Acoust. Soc. Amer.* <u>71</u>, 588 (abstract).

Strelioff, D. and Flock, Å. (1982). Mechanical properties of hair bundles of receptor cells in the guinea pig cochlea. *Soc. Neurosci.* Abst. 8.

Tilney, L.G. and Saunders, J.C. (1982). Actin filaments, stereocilia and hair cells of the bird cochlea I. The length, number, width and distribution of stereocilia of each hair cell is related to the position of the hair cell on the cochlea. (in press).

Tilney, L.G., Saunders, J.C., Egelman, E. and De Rosier, D.J. (1982). Changes in the organization of actin filaments in the stereocilia of noise-damaged lizard cochlea. *Hearing Res.* <u>7</u>, 181-197.

Wever, E.G. (1983). *The amphibian ear* (in press).

Wright, A. (1981). Scanning electron microscopy of the human cochlea - the organ of Corti. *Arch. Oto-Rhino-Laryngol.* <u>230</u>, pp. 11-19.

Zwislocki, J.J. (1981). Sound analysis in the ear: a history of discoveries. *Amer. Scientist* <u>69</u>, pp. 184-192.

Section VI

Special topics

THERMAL AND QUANTUM NOISE IN THE INNER EAR

William Bialek

Department of Biophysics and Medical Physics, and
Division of Biology and Medicine, Lawrence Berkeley Laboratory,
University of California, Berkeley
Berkeley, California 94720 U.S.A.

ABSTRACT

General theories of thermal and quantum fluctuations are used to calculate the levels of noise in models of inner ear mechanics. In each case considered, the calculated levels of thermal noise are much too large to be consistent with the detection of sub-angstrom motions at the threshold of hearing. Quantum noise levels are comparable to these threshold signals, implying that the inner ear is not a classical system. Some implications of these results for theories of hearing are noted.

1. INTRODUCTION

In the classical view of the cochlea, the stereocilia of the receptor cells

respond passively to forces which result from basilar membrane displacement.

In recent years, numerous revisions of this classical view have been suggested;

most of these suggestions were presented as possible explanations for the

remarkable filtering abilities of the mammalian cochlea and other inner ear

organs (cf. Lewis *et al.* (1983) for review).

In this paper I shall discuss some implications of another remarkable ability

of the inner ear, namely its detection of small signals. The most recent

measurements in the cat cochlea find $10^{-10}m$ motions of the basilar membrane at

sound pressure levels 30 dB above the threshold for a reliable behavioral

response of the animal (Khanna and Leonard, 1982). This suggests that the

displacements at threshold are $10^{-11}m$ or less, and this is consistent with

results (Peake and Ling, 1980) from the alligator lizard basilar papilla

(cf. Bialek and Schweitzer (1983a)). Similarly, neurons from the saccule of

the white-lipped frog display clear responses to $10^{-11}m$ vibrations of the whole frog (Lewis and Narins, 1981).

Can the classical model, or any of its modern variants, account for the detection of $10^{-11}m$ displacements? Three investigators, adopting different theoretical methods, have reached widely diverging conclusions: De Vries (1956) and Flerov (1976) argue that the levels of thermal noise expected in the classical model are much larger than the threshold signals, while Harris (1968) suggests that the Brownian motion of the basilar membrane could be well below the threshold displacement. Allan Schweitzer and I (Bialek and Schweitzer, 1983a; 1983b), in a more systematic approach to the problems of noise in the inner ear, find that thermal noise levels are much greater than the threshold signals, and quantum noise levels are (perhaps surprisingly) significant.

2. THERMAL NOISE AT THE STEREOCILIUM

The detector elements of the inner ear are the stereocilia, which are roughly cylindrical objects of length $L = 4\mu m$ and radius $R = 50\ nm$; they consist of a crosslinked bundle of actin filaments (Flock and Cheung, 1977). From these facts we can estimate the mechanical properties of the cilium to be expected if the system is mechanically passive. All proteins, including actin, have a density of $\rho = 1.3\ gm\text{-}cm^{-3}$ and a Young's modulus of $Y \leq 2x10^{10}Nt\text{-}m^{-2}$ (Karplus and McCammon, 1979). Thus a single stereocilium will have a mass $m=\pi\rho R^2 L=4x$ $10^{-17}kg$. If the stereocilium is clamped at its base and free at its tip, then it will move in a cantilevered mode and have a stiffness $\kappa=3\pi YR^4/16L^3 \leq 10^{-3}$ $Nt\text{-}m^{-1}$. Indeed the cilia of many receptor cells are free-standing, and hinging of the base will only decrease the stiffness, so that our upper bound is correct.

The upper bound on the stiffness of a stereocilium determines the thermal

noise displacements (Brownian motion) of the cilium which we would see if we
made broad-band measurements of its motion. The result (Landau and Lifshitz,
1977) is that the mean-square displacement $<(\delta x)^2>=k_BT/\kappa$, where $k_B=1.36x10^{-23}$
$J-K^{-1}$ is Boltzmann's constant and $T=300K$ is the absolute temperature. Thus
$\delta x_{rms}=(k_BT/\kappa)^{1/2}\geq 2x10^{-9}m$, which is 40 dB above the threshold displacements in
the inner ear.

To calculate the spectral density S_x of stereocilium Brownian motion, we need
an estimate of its damping coefficient. An order of magnitude estimate may
be obtained from the same hydrodynamic considerations which arise in the
analysis of ciliary beating (Lighthill, 1975); for an object with the dimensions given above we obtain $\gamma=10^{-10}Nt-s-m^{-1}$, assuming that the viscosity of
fluid surrounding the cilium is close to that of water. From the fluctuation-
dissipation theorem (Landau and Lifshitz, 1977),

$$S_x(\omega) = \frac{\gamma k_BT/\pi}{(\kappa-m\omega^2)^2+(\gamma\omega)^2} \, , \tag{1}$$

where the root-mean-square fluctuations in a narrow bandwidth Δf are $\delta x_{rms}=$
$(4\pi S_x\Delta f)^{1/2}$. From the parameter estimates above, $\delta x_{rms}>1.3x10^{-12}m(\Delta f/1Hz)^{1/2}$
for frequencies in the normal auditory range. If we are to detect $10^{-11}m$ dis-
placements of single stereocilia, then the hair cell must possess a filter
with a bandwidth of *50 Hz* or less.

It might be supposed that much larger detection bandwidths could be tolerated
by averaging over the many stereocilia on each hair cell; this is not the
case. At, for example, *1 kHz*, fluid motion extends around the stereocilium
through a boundary layer of depth = *20*μm (for the viscosity of water;
Landau and Lifshitz, 1959), and objects within this layer--such as cilia on a
single receptor cell, or even nearby cells--will be coupled through the
viscosity of the fluid. Systems which are viscously coupled exhibit
correlated Brownian motions (Landau and Lifshitz, 1977), and hence there will

be little noise reduction upon averaging.

3. THERMAL NOISE AT THE BASILAR MEMBRANE

It might be possible to overcome stereocilium Brownian motion by amplifying
the motion of the basilar membrane, but this will be disastrous if the
Brownian motion of the membrane itself is significant. To estimate the
Brownian motion of the membrane, we write the energy of the system in terms of
the position and velocity of the membrane at each point in the cochlea, and
then apply the Boltzmann distribution.

In conventional models of membrane mechanics (de Boer 1970; Lighthill, 1980;
Lewis *et al.*, 1983), the energy consists of three components: the kinetic
energy of the basilar membrane, the kinetic energy of the cochlear fluids, and
the potential energy of the membrane. Considering the kinetic energy terms,
the basilar membrane velocity $\nu(x)$ at the point x along the cochlea has the
spatial Fourier representation

$$\nu(x) = \int \frac{dk}{2\pi} e^{ikx} V(k),$$ (2)

so that the kinetic energy of the membrane

$$K.E. (membrane) = (mb/2)\int dx \nu^2(x) = (mb/2)\int \frac{dk}{2\pi} |V(k)|^2,$$ (3)

where m and $b = 100\mu m$ are the mass per unit area and width of the membrane,
respectively. The kinetic energy of the fluid can be written

$$K.E. (fluid) = (b/2)\int \frac{dk}{2\pi} \rho Q^{-1}(k) |V(k)|^2,$$ (4)

where $\rho = 1 \ gm\text{-}cm^{-3}$ is the fluid density and $Q(k)$ depends on the assumed
geometry of the model (Lighthill, 1980). Thus, if the fluid motion is primarily one-dimensional, $Q(k) = hk^2$, where $h = 0.1cm$ is the effective height of
the cochlear chambers, while if the motion is two-dimensional $Q(k) = k\tanh(kh)$.
The total kinetic energy is therefore

$$K.E.\,(total) = (b/2)\int \frac{dk}{2\pi}\,[m+\rho Q^{-1}(k)]\,|V(k)|^2.\tag{5}$$

According to the Boltzmann distribution, the probability of any particular configuration of the system is $e^{-E/k_B T}$, where E is the energy of the configuration (Landau and Lifshitz, 1977). Applying Eq. (5) for the energy, the $V(k)$ are Gaussian random variables; the variances are given by

$$<V(k)V(k')> = \frac{k_B T\delta(k+k')}{b[m+\rho Q^{-1}(k)]}.\tag{6}$$

so that

$$<\nu^2(x)> = (k_B T/mb)\int \frac{dk}{2\pi}\,\frac{Q(k)}{Q(k)+\rho/m}\tag{7}$$

For $Q(k)$ given above the integral may be evaluated as

$$<\nu^2(x)> = (k_B T/2bh^2\rho)\,(\rho h/m)^{3/2} \qquad \rho h \ll m,\tag{8a}$$

$$= (k_B T/bh^2\rho)\,(\rho h/m)^2 \qquad \rho h \gg m.\tag{8b}$$

Current models often assume $m = \rho h$, so that the thermal fluctuations in basilar membrane velocity will be $\delta\nu_{rms} = (k_B T/bh^2\rho)^{1/2} = 2\times10^{-5}$ cm-sec^{-1}.

At a frequency of 1 kHz, these velocity fluctuations are equivalent to displacement fluctuations of $= 10^{-10}m$, or more than 20 dB above threshold; the problem is correspondingly worse at higher frequencies. It may be shown that the correlation length for these fluctuations is $= h$, so that no reasonable spatial averaging could reduce their effect.

The same reasoning may be applied to the potential energy, which is

$$P.E.\,(membrane) = \frac{b}{2}\int dx C(x)z^2(x),\tag{9}$$

where $z(x)$ and $C(x)$ are the displacement and stiffness per unit area of the membrane, respectively. In analogy to Eq. (6), we find

$$<z(x)z(x')> = k_B T C^{-1}(x)\delta(x-x').\tag{10}$$

If we average over a region of length d, we will see displacement fluctuations of $\delta x_{rms} = [k_B T/bdC(x)]^{1/2}$, and C may be determined from the resonance condition $\omega_o^2 m = C$. Thus, at the position in the cochlea corresponding to $\omega_o =$ $2\pi(1kHz)$, and with $m = \rho h$ from above, a hair cell of diameter $d = 10\mu m$ will "see" a basilar membrane Brownian motion $\delta x_{rms} = 3x10^{-10}m$, or about 30 dB above threshold. Whether we measure displacement or velocity, the thermal noise at the basilar membrane is large--too large to allow significant amplification of the membrane motion (including through feedback; cf. Gold (1948)) without making a serious problem much worse. The need to reduce the effects of this noise argues strongly for a filtering process subsequent to basilar membrane mechanics.

4. QUANTUM NOISE AT THE STEREOCILIUM

If we imagine a fictitious inner ear operating at absolute zero, there would be no thermal noise, but there are still quantum limits to the signals which can be reliably detected. In particular, if we try to follow the amplitude and phase of stereocilium motion--as the hair cell does by producing an AC receptor potential--then we will see a displacement noise, or zero-point motion (Braginsky et al.,(1980), $<(\delta x)^2> = h/2(m\kappa)^{1/2}$, where m and κ are the mass and stiffness of the stereocilium, respectively, and $h=1.054x10^{-34}J\text{-}sec$ is Planck's constant. With the estimates of these parameters given above, the zero-point motion of the stereocilium is greater than $10^{-12}m$, or within an order of magnitude of threshold. This rough calculation demonstrates that the inner ear, near threshold, operates in a regime where quantum mechanical effects are non-negligible.

Quantum mechanics also places limits on the performance of amplifiers used in the measurement process: any linear amplifier must contribute some minimum excess noise to the signals which it amplifies (Caves, 1982). If we are to make measurements near the limit imposed by zero-point motion, then our amplifier must contribute no more than this minimum noise. Apparently the amplification processes occuring within the hair cells of the inner ear are "perfect" in this quantum mechanical sense.

Is it plausible to suppose that "perfect" amplification occurs in a biological system? A number of biological systems are well described by theories in which quantum mechanical effects are explicit and significant (e.g. DeVault, 1980). By extending these theories it is possible to "build" a perfect amplifier out of the known properties of biological molecules (Bialek and Schweitzer, 1983b). We therefore should not be surprised by the significance of quantum effects in a system which, after all, operates at a sub-molecular scale. The fact of our surprise remains, however.

Acknowlegements

I thank Allan Schweitzer for permission to quote from our unpublished results, Professor G. Zweig for helpful discussions, and Professor A. Bearden for a productive research environment. This work was supported by the National Science Foundation Biophysics (PCM 78-22245) and Pre-Doctoral Fellowship programs, and by the Office of Basic Energy Sciences, Office of Energy Research, U.S. Department of Energy, under contract No. DE-AC-03-76SF00098.

REFERENCES

Bialek, W. and Schweitzer, A. (1983a). Thermal noise and the auditory receptor cell. In preparation.
Bialek, W. and Schweitzer, A. (1983b). Quantum noise and the molecular basis of auditory detection. In preparation.
Braginsky, V.B., Voronstsov, Y.I., and Thorne, K.S. (1980). Quantum non-demolition measurements. *Science 209*, 547-557.
Caves, C. M. (1982). Quantum limits on noise in linear amplifiers. *Phys. Rev. D 26*, 1817-1839.
De Boer, E. (1980). Auditory physics. Physical principles in hearing theory. I. *Phys. Repts. 62*, 87-174.
DeVault, D. C. (1980). Quantum mechanical tunneling in biological systems. *Quart. Rev. Biophys. 13*, 387-564.
De Vries, Hl. (1956). Physical aspects of the sense organs. *Prog. Biophys. Biophys. Chem. 6*, 207-264.

Flerov, M.N. (1976). Thermal noise of hair cells in the organ of Corti. *Biofizika* 6, 1092–1096.

Flock, A. and Cheung, H.C. (1977). Actin filaments in sensory hairs of inner ear receptor cells. *J. Cell. Biol.* 75, 339–343.

Gold, T. (1948). Hearing. II. The physical basis of the action of the cochlea. *Proc. Roy. Soc. Edinb.* B135, 492–498.

Harris, G.G. (1968). Brownian motion in the cochlear partition. *J. Acoust. Soc. Am.* 44, 176–186.

Karplus, M. and McCammon, J.A. (1980). The internal dynamics of globular proteins. *C.R.C. Crit. Rev. Biochem.* 9, 293–349.

Khanna, S.M. and Leonard, D.G.B. (1982). Basilar membrane tuning in the cat cochlea. *Science* 215, 305–306.

Landau, L. and Lifshitz, E.M. (1977). *Statistical Physics* (Pergamon, Oxford).

Landau, L. and Lifshitz, E.M. (1959). *Fluid Mechanics* (Pergamon, Oxford).

Lewis, E.R., Leverenz, E.L., and Bialek, W.S. (1983). The vertebrate inner ear. To appear in *C.R.C. Rev. Biomed. Eng.*.

Lewis, E.R. and Narins, P.M. (1981). Seismic sensitivity in VIIIth nerve afferent fibers of the white-lipped frog. *Soc. Neurosci. Abs.* 7, 148.

Lighthill, J. (1975). *Mathematical Biofluiddynamics* (SIAM, Philadelphia).

Lighthill, J. (1980). Energy flow in the cochlea. *J. Fluid Mech.* 106, 149–213.

Peake, W.T. and Ling, A.R. Jr. (1980). Basilar membrane motion in the alligator lizard: Its relation to tonotopic organization and frequency selectivity. *J. Acoust. Soc. Am.* 67, 1736–1745.

A HAIR CELL MODEL OF NEURAL RESPONSE

J. B. Allen

Bell Laboratories
Murray Hill, New Jersey 07974

ABSTRACT

In this paper we present a model of neural response to pure tones and tone bursts. The model data fit the neural data over a large range of frequencies and levels. Some exceptions to this fit are discussed. The model is based on the Davis model of transduction which describes the cell receptor potential given cilia displacements. We augment the Davis model by assuming, as was done in the Schroeder-Hall model, that the neural response is proportional to the cilia current. We further assume that the current is low pass filtered in order to account for loss of the phase locked response for frequencies above 5 kHz.

1. INTRODUCTION

In recent years an increased emphasis has been placed on improved understanding of cochlear hair cell transduction. In this period a great deal of experimental data has demonstrated the validity of the Davis model of transduction. In 1958 Davis proposed that basilar membrane motion produces hair cell resistance charges, thereby modulating the receptor potential within the body of the hair cell. This idea has now been tested in many laboratories, and in all cases the model appears to be in agreement with the experimental data.

However the transformation that leads to the probability of firing at the neuron level has not been as successfully explained, although it has been attempted [Schroeder and Hall, (1974); Smith and Zwislocki (1975); Ross, (1982)]. In the opinion of this author, one common flaw with each of these attempts was that they were not predicated on the Davis model. In the presentation developed here we attempt to build on the basic Davis model, which seems to accurately model the receptor potential. We then study the type of transformation that is required to transform the receptor potential into a neural response. We next compare the model neural data to experimental data, discuss the weakness of the model, and discuss the the problems of physiological correlates. This final question relates to a basic unsolved problem in neurobiology, namely the question of the release of vesicles at the synaptic site. It is presently believed that calcium is responsible for the fusion of vesicles at the cell wall. Vesical fusion then leads to the release of transmitter substance into the synapse. This general mechanism is therefore likely to be involved in the signal path between the hair cell receptor potential and the neuron.

2. THE DAVIS MODEL

The basic model, proposed in 1958 by Davis, has evolved into the more complete model shown in Fig. 1, which is reproduced from Weiss et al. (1974). Because the model of Fig. 1 has evolved beyond the original Davis proposal, I shall refer to it here as the variable resistance model. This model, which describes the receptor potential in terms of the displacement of the cilia, has been directly tested in several ways.

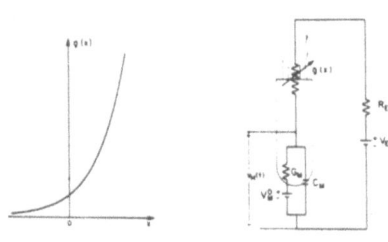

Fig. 1. This figure has been reproduced from Weiss et al. (1974). This model differs from the original Davis model in that membrane capacitance and conductance have been included. The cilia conductance g(x) is assumed to form a half-wave rectifier as a function of the cilia displacement x, which in turn is believed to result from the shear between the tectorial membrane and the top of the hair cell (reticular lamina).

a. Receptor potential

The receptor potential has been experimentally measured by Flock (1971), Weiss et al. (1974), Russell and Sellick (1978), Dallos (1975), Hudspeth and Corey (1977), and Holton (1981). In Fig. 2 we see recent receptor potential measurements for four different cells (Holton, 1981) driven at each units characteristic frequency (CF). Two features are notable: *First,* the component in the response at the stimulus frequency decreases with increasing stimulus frequency. This effect is accounted for in the variable resistance model by the low-pass filter resulting from the cell capacitance which is in series with the cilia resistance. *Second,* the membrane voltage (envelope) increases monotonically when the stimulus is applied, in a manner predicted by the variable resistance model. (This charging curve will be compared to the neural response which is known to be quite different.) Thus the waveform of the receptor potential (Fig.2) seems to be, at least qualitatively, in agreement with the model of Fig. 1.

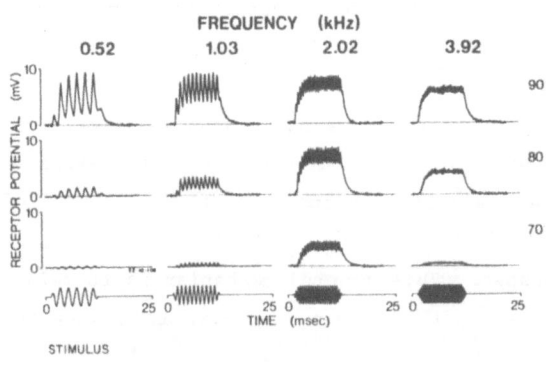

Fig 2. The data shown here (from Holton, 1981) are receptor potential measurements from alligator lizard hair cells. In this figure the stimulus was a tone bursts (of various frequencies and levels). Note how the potential increases, when the tone is turned on, with a constant of about 1 ms (one full period of 1.03 kHz). Note also the similar discharge time after the tone has been turned off.

b. *Level dependence*

A third important characteristic of the cell, beside the rise time and the frequency response, is the level of dependence. This level dependence can be measured several different ways, however, basically it is the characterization of the response of the cell as a function of the excitation level. Frequently this data is presented as a family of curves with frequency being the parameter (Russell and Sellick, 1978).

Available level data frequently does not agree with the variable resistance model in that the DC saturation value for the receptor potential can be frequency dependent (relative to the CF). According to the variable resistance model, the time average receptor potential should saturate at a voltage which is independent of frequency.

c. *Resistance measurements*

A recent important check on the variable resistance model was made by Hudspeth and Corey (1977), and some of their results are shown in Fig. 3. Since their data are for bullfrog sacculus hair cells, the absolute sensitivities may not be representative of the mammalian auditory system.

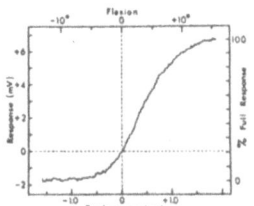

Fig. 3. We show here the input-output relationship for a hair cell. The curve shows the potential change from the resting potential (-58 mV) as a function of the displacement of the hair bundle's tip by a 10-Hz triangle wave stimulus. An alternate abscissa represents the estimated angle of flexion of the 9 μm-long hair bundle on the assumption that it pivots at its base.

3. NEURAL RESPONSE MODEL

The receptor potential is merely an intermediate step between the cilia displacement and the neural response, and a great deal more neural data is available in the literature than receptor potential data. The modeling position represented here is that an active linear transformation (a linear filter) will transform the receptor potential data into the neural response.

Smith and Zwislocki (1975) also determined that the relation between the receptor potential and the neural response seemed consistent with a linear transformation. Their results are based on scaling relationships that they observed from a series of experiments using tones having incremental steps. Thus the model proposed here is in agreement with their results since we specifically assume a linear transformation.

a. *Previous neural models*

The most widely acknowledged neural model to date has been the model of

196

Schroeder and Hall (1974). One notable aspect of the Schroeder-Hall model
was their attempt at giving a physiological basis for the model. Unfortunately this model had several deficiencies. *First*, it is lacking a
description of the hair cell receptor potential, such as described by the
Davis variable resistance model. As a result, it is very difficult to relate
the model to this physically measurable quantity. *Second*, their maximum rate-level curve does not saturate for large levels. This deficiency is
easily corrected by simple modifications to the model, thus it does not
appear to be a major limitation. *Third*, the model does not exhibit a
loss of phase locking for frequencies above 5 kHz, as is commonly observed in
neural data. Again, perhaps this deficiency could be easily corrected by a
modification to the original model.

The Schroeder-Hall model may be transformed into the form of the Davis model,
by the use of the Thevenin equivalence theorem. However when this is done,
the voltage source for the Schroeder-Hall model is signal dependent, whereas
the voltage source for the Davis model is constant. Thus, the two models are
formally different. The Davis and the Schroeder-Hall models are similar if
one makes three changes, or identifications. *First* the voltage source
in the Schroeder-Hall model, after the Thevenin transformation, must be set
to a constant, to make it equivalent to the Davis model. *Second* we identify the neural response with the current through r(t), as is assumed by the
Schroeder-Hall model. *Thirdly*, we must low-pass filter the current to
account for the loss of a phase locked response in the neural data. If we
make these three changes, we may combine the success of the two models.

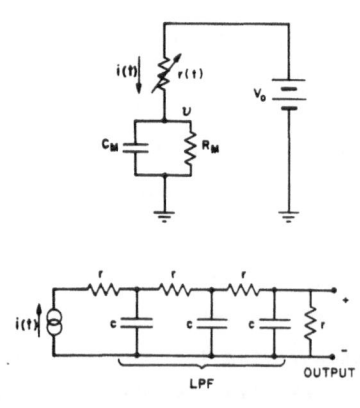

*Fig. 4. This figure shows the model as pro-
posed in this paper. We start with a simplified version of the Davis model for the cell
receptor poetential. We then take the current
as the output, rather than the voltage. The
current and voltage are related by the linear
transformation of Eq. 1. Finally, in order
that the response not be phase-locked at high
frequencies, we low pass filter the current in
order to produce the model response. The low-
pass filter is shown here as a diffusion
transmission line. Some justifications exist
for this particular choice of low-pass filter.
However improved measurements are needed to
determine the true type of filtering which
removes the phase locked component above 4 or
5 kHz.*

b. *A composite model*

In Fig. 4 we show the composite model as described above. The current in the
variable resistance model is taken as the input to a low-pass filter labeled

LPF. The output of the low-pass filter must show no phase locked component for frequencies above 5 kHz since stimulus frequencies above this value are never seen in neural data.

Next it is useful to recognize that the current may be computed from the receptor potential by linear operations since the current through r(t) is equal to the sum of the currents through the capacitor and the cell leakage resistance. Thus

$$I(s) = V(s)(\frac{1}{R_m} + sC_m) \tag{1}$$

where $s = \sigma + i\omega$ is the Laplace transform complex frequency variable, Rm is the cell leakage resistance, and Cm is the total cell capacitance (Fig. 1).

According to this model it is possible to compute the model neural response . by an *active* linear transformation of the hair cell receptor potential We now present data that support this conclusion.

c. Model responses to pure tones

The data being modeled in this section are stimulus locked single tone histogram data of D. Johnson (1974). Pure tones were presented to the cat while single unit spikes times were recorded modulo the tone period. A histogram was then made of the probability of firing relative to the stimulus period.

The model and neural results are compared in Fig. 5. The first two columns of this figure give the 200 Hz data at 10 dB levels, the second two columns are for 1800 Hz, and the last two columns give 3.4 kHz data. At low levels - the unit fires at the spontaneous rate with equal probability over the inter val. As the level increases above the spontaneous rate the firing rate develops a phase locked sinusoidal component. For frequencies above 4 or 5 kHz, a phase locked component is never observed.

As the level is further increased the observed response depends on frequency For frequencies near 4kHz the response remains sinusoidal. For lower frequencies the firing probability estimate appears half wave rectified, except at low frequencies where it has the characteristic shape seen in the bottom-most panel of the first column.

The model results seen in the corresponding (second) column match the experimental data reasonably faithfully. The bottom-most panel in each model column gives the peak "rate" level curve as obtained from the model. These peak rate curves have been normalized to the maximum rate and are very similar to corresponding experimental data.

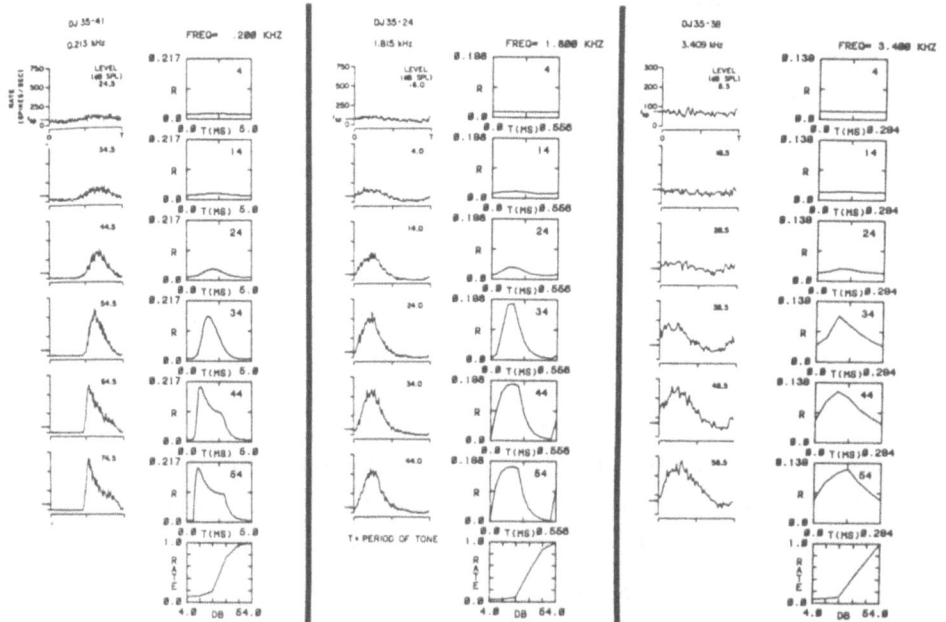

Fig. 5. *In this figure we compare model calculations to the single-tone neural data of Johnson (1974). The first column is the neural data, as a function of level, in 10 dB steps, for a frequency of 213 Hz. The ordinate is the estimated probability of firing, as a function of the tone period. All panels display one period of the stimulus independent of frequency. The second column is the model result for 200 Hz. The bottom-most panel gives the maximum model output (firing rate) as a function of input level. The model rate-level curve is very similar to neural rate-level curves. Columns three and four give similar data for a frequency of 1.8 kHz. Columns five and six are for a frequency of 3.4 kHz.*

In order to qualitatively understand the low-frequency high-level curve, it is useful to think of the cilia resistance of the model r(t) as being switched between one of two possible resistances rLO and rHI, depending on the sign of the cilia excitation. When the cell resistance switches to rLO, the receptor capacitance begins to charge, and the cell voltage changes exponentially to its new steady state potential. The current through r will increase dramatically at first because the voltage drop across r is large. As the voltage v increases toward VE the voltage across r decreases, thus the current decreases.

d. *Model response to tone bursts*

Next we look at the case of a tone burst stimulus since it is a more compli-

cated stimulus. In Fig. 6 we show neural data collected by the author (from
cat). The tone burst was presented at the unit's CF (110 Hz). The unit had
a normal CF threshold for this frequency range and had a very broad flat
response near the CF. The histogram bin width was 50 μsec; however, for
plotting purposes the data was averaged by an eight sample long rectangular
window and was decimated eight to one. The lowest panel displays the maximum
instantaneous rate as a function of the sound level. The number in the upper
right hand corner of each panel is the sound pressure in dB re 20 μPa. The
decrease in rate at 75 dB is due to the first peak suppressing the response
of the second peak, thus decreasing the maximum rate at the second peak.
Such nonmonotonic behavior seems to be relatively common at low frequencies.

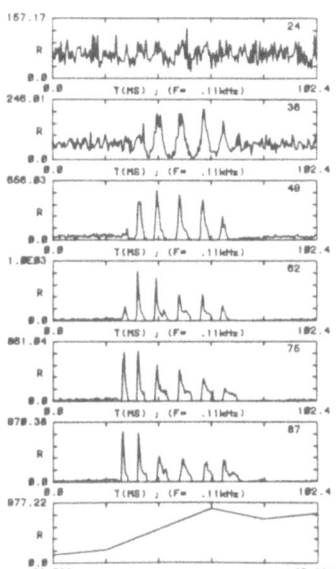

*Fig. 6. In this figure we show neural tone burst
data measured in cat by the author using methods
described in Allen (1983). The stimulus frequency was 110 Hz, and the sound level went from 24
dB-SPL (dB re 20 μPa) to 87 dB-SPL. The bottom-
most panel shows the maximum rate as a function
of level of the tone. The dip in rate at 75 dB-
SPL is due to the increase in firing of the first
peak which had a suppressive effect on the second
peak, thereby reducing the maximum rate. As the
pressure increased further, the first peak rate
increased to the previous maximum of the second
peak. Note also the characteristic waveform
which, at high levels, shows a peak firing
response, which then decreases to a plateau. For
one half cycle the unit fails to fire. Note also
that the maximum firing rate on the 5th cycle is
about 1/3 that of the first cycle for the highest
input levels.*

In Fig. 7 we show the hair cell model response to
the identical tone burst. The left hand column
gives the model cilia resistance values on a log scale. Note that for this
figure the intensity changes by 10 dB for each panel whereas in the previous
figure the sound intensity changed by 12.7 dB between panels. For Fig. 7 the
middle column gives the model receptor potential (compare this to Fig. 2),
while the right hand column shows the low-pass filtered current, which
according to the model represents the probability of neural firing. In the
bottom-most panel of the right hand column we show the maximum model firing
rate as a function of level. Note the decrease in firing rate due to the
suppression of the second peak by the first at 54 dB. Also note the
similarity between the maximum rate-level curves of Figs. 6 and 7.

Two differences are seen between the neural and model data. First the maximum firing rates for the third, fourth and fifth period are about half the
maximum for the experimental data. In the case of the model these maxima

only drop about 40%. Second the time constant for recovery of the spontaneous rate for the neural data (Fig. 6 is about 10 ms, while the model recovery time (Fig. 7) is somewhat shorter. These two failures seem to be similar in their time constant and may represent a common model failure, or model parameter error. In general the agreement seems excellent.

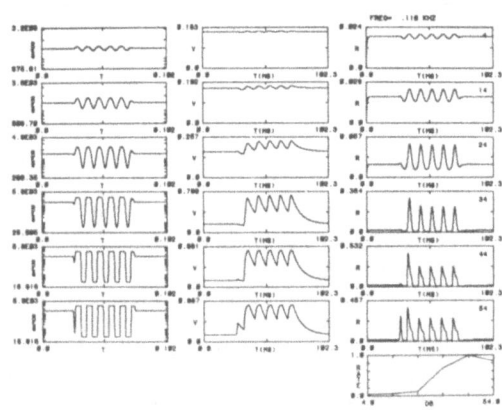

Fig. 7. These model results should be compared to the previous figure which gives neural data under the same conditions. For the model we show three columns. The first is the cilia resistance as a function of time. The resistance is plotted on a log-ordinate scale. In the second column we show the model hair cell receptor potential. In the last (right-most) column we show the model neural response, as computed from the model of Fig. 6.

In Figs. 8 and 9 we compare similar data (Fig. 8 is neural data from a cat unit having a CF of 4.85 kHz, while Fig. 9 is the corresponding hair cell model result). The ordinate, as before, gives the instantaneous firing rate. The bottom panel gives the maximum firing rate over the stimulus period as a function of level. Again the agreement seems excellent.

In Fig. 9, as in Fig. 7, we show the model cilia resistance, receptor potential, and firing rate, as a function of stimulus level. Note that the rise time for the receptor potential decreases as the level increases, as would be expected from the model for smaller charging resistances at large stimulus levels. Note also that the same effect may be seen in the neural data in Fig. 8 where the leading edge becomes relatively sharp at large sound levels (73 dB).

4. DISCUSSION

A most important aspect of modeling is identifying the model features with measurable physical (in this case biophysiological) quantities. In the case at hand we do not have sufficient experimental data to make definitive associations. On the other hand there is a large body of literature on the chemical synapse since it is the object of intense research because of its pervasiveness in neural systems. If we assume that cochlear hair cells are similar to other biophysical systems, then it is possible to speculate on a physical realization of the hair cell model.

According to the present view of chemical synaptic transmission (Llinas et al., 1981) the following chain of events is assumed to occur: First the receptor potential controls the opening of calcium channels in the cell wall at or near the synapse. These ions then diffuse to the pre-synaptic membrane where they cause vesicles to fuse with the cell wall, thereby dumping their contents (transmitter substance) into the synaptic cleft. Either calcium currents or vesicle depletion might account for the initial burst in spike activity, and the slow decay in the firing rate for long times (10-20 ms) [e.g., transformation given by Eq. (1).] The low-pass filter of the model presented here might be due to the diffusion of the calcium ions from the cell wall to the synapse where the vesicles reside.

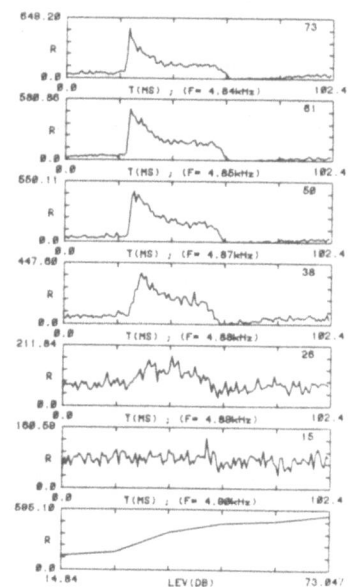

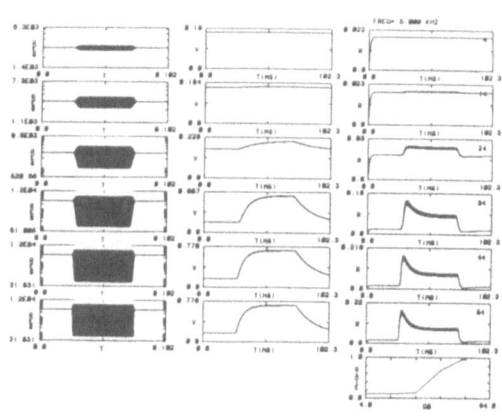

Fig. 8. We show here tone burst data (from cat) for a stimulus frequency of 4.8 kHz. Note how the leading edge of the tone burst response becomes very steep at high levels. Note also that no 4.8 kHz component is seen in the neural response. After the tone response has been turned off, the spontaneous response slowly recovers with a time constant of about 20 ms.

Fig. 9. The model conditions for this figure are nearly identical to those of the neural data in Fig. 10. As in Fig. 9 the first column shows the cilia resistance on a log scale. The second column gives the model receptor potential. The third column is the model neural response. After the tone has been turned off, the recovery is too rapid, relative to the neural data, however the overall shape and response seems quite reasonable. The rate level function is typical relative to neural data, but the phase locked component is too large, indicating that the simple RC filter used here was not sufficient to remove the higher frequency components.

REFERENCES

Allen, J. B. (1983). "Magnitude and phase frequency response to single tones in the auditory nerve," to appear, June, *Jol. Acoust. Soc. Am.*

Davis, H. (1958). "A Mechano-Electrical Theory of Cochlear Action." Ann. Otol. Rhinol. *Laryngol. 67*, 789-801.

Flock, A. (1971). "Sensory Transduction in Hair Cells," in Handbook of Sensory Physiology, Ed. by W. R. Lowenstein (Springer-Verlag, New York), Vol. I, pp. 396-441.

Holton, T. (1981). "Mechanoelectric Transduction by Cochlear Receptor Cells," Ph.D. thesis, Mass. Inst. of Tech., Thesis Advisor Prof. T. Weiss, Dept. of EE.

Hudspeth, J., and Corey, D. P. (1977). "Sensitivity, polarity and conductance change in the response of vertebrate hair cells to controlled mechanical stimuli" *Proc. Natl. Acad. Sc. USA, 74*, pp. 2407-2411.

Johnson, D. H. (1974). "The Response of Single Auditory-Nerve fibers in the cat to single tones: Synchrony and average discharge rate," Ph.D. thesis, MIT.

Llinas, R. R., Steinberg, I. Z., Walton, K. (1981). "Relationship between Presynaptic Calcium current and Postsynaptic Potential in Squid Giant Synapse" *Biophysical Journal, 33*, 323-352.

Ross, S. (1982). "A model of the hair cell-primary fiber complex." *J. Acoust. Soc. Am. 71, 926-941.*

Russell, I. M. and Sellick, P. M. (1978). "Intracellular Studies of hair cells in the mammalian cochlea" *J. Physiol 284*, 261-290.

Schroeder, M. R., Hall, J. L. (1974). "Model for mechanical to neural transduction in the auditory receptor," *J. Acoust. Soc. Am. 55* pp. 1055-1060.

Smith, R. L. and Zwislocki, J. J. (1975). "Short-term Adaption and Incremental Responses of Single Auditory-Nerve Fibers," *Biol. Cybernetics 17*, 169-182.

Weiss, T. F., Mulroy, M. J., Altman, D. W. (1974). "Intracellular response to acoustic clicks in the inner ear of the alligator lizard," *J. Acoust. Soc. Am. 55*, 606-619.

STATIC POINT-LOAD MEASUREMENTS OF BASILAR MEMBRANE COMPLIANCE

C.E. Miller

*Dept. of Mechanical Engineering, Stanford University
Stanford, CA 94305*

ABSTRACT

Point-load measurements of basilar membrane compliance in the guinea pig were made with the intent of providing insight into the membrane's structural behavior and parameters for use in mathematical modeling. Selected points on a radial line across the basilar membrane were statically displaced in one-half micron steps and the force required to maintain each increment of displacement recorded. Curves of force versus deflection were constructed for each point tested and appear trilinear in the 0-8 micron range of deflection. The slope of these curves in the steepest region was then plotted as a function of normalized radial distance. Consistent among all tests is a comparatively stiff arcuate zone. Calculations of point-load compliance and volume compliance were compared to results obtained by other investigators.

1. INTRODUCTION

There is little doubt that the basilar membrane (BM) is the dominant elastic

element of the cochlear partition. An accurate determination of its structural

properties is thus important for the comparison of cochlear models with the

physical system. With the demonstration that calculations can be carried out

for three-dimensional models (Taber and Steele (1981)), the knowledge of the

variation of BM compliance with radial distance becomes necessary for the de-

termination of BM motion. To date, however, experimental data on this

variation is sparse. The first experiments were done by Békésy, who measured

volume compliance in various mammals and also point elasticity in the human.

The age and physiological condition of the samples in both cases is unknown,

though, and the results obtained in the second case, by loading with a series

of small hairs, are limited by the discrete nature of the values for force.

Dancer and Franke (1980) estimated the volume compliance of the live guinea

pig to be one-fifth of Békésy's *post mortem* measurements at a point 2 mm from
the basal end. This would indicate an *in vivo* value of about .04 mm^4/N at the
2 mm point. Gummer et.al. (1981), in the *post mortem* guinea pig cochlea,
measured the slope of the dynamic force-deflection curve as a function of
static BM deflection for radially central locations in the basal region. Their
results ranged from 1.22 - 3.57 m/N.

In order to obtain a more complete assessment of BM parameters, the following
experiments were performed to measure the point-load compliance of the BM as a
function of radial distance. The guinea pig was used because of the experimental results already available, but it was necessary to work with the excised
cochlea because of the requirement of perpendicular access to the BM.

2. MATERIALS AND METHODS

a. Specimen preparation

Guinea pigs weighing between 250 and 300 grams were sacrificed with an overdose
of Nembutal. The ear bullae were quickly removed and opened and the exposed
cochleae embedded apex downward in dental cement. Scalpel and pick were used
to carefully remove a section of bone of scala tympani to allow access for
viewing and access by the force transducer. Care was taken to avoid contact
with the BM and to preserve its attachments. The position exposed varied from
1.0 to 2.5 mm from the basal end. A layer of perilymph was left above the
membrane and a fluid reservoir created by surrounding the specimen with gauze
soaked with Ringer's solution. A 6-7 micron layer of fluid was then maintained at all times above the membrane. The specimen was firmly clamped, scala
tympani side of BM upward, in a frame providing three rotational degrees of
freedom.

b.   Measurement technique

The system used to measure force involved a servo accelerometer (Systron Donner Inc., Concord, Calif., model 4311-1-X15) of the type generally used in inertial guidance systems. The seismic mass, mounted on an arm of the accelerometer's pendulous element, was removed and a glass needle with tip diameter of 12.5 microns glued in its place. A force applied to the arm causes a movement of the arm which is detected by a position sensor. An error signal adjusts the current to a torque-restoring coil which returns the arm to its original position. The voltage drop across a resistor in series with the coil can be measured and is proportional to the applied force.

The force transducer was mounted in a micromanipulator providing three degrees of translational freedom. The specimen was placed below the glass needle and oriented with the needle perpendicular to the plane of the BM. A Wild stereo-microscope on 50X magnification was used to verify all positionings. The trans-ducer/needle was lowered to the BM at a point just radial to the edge of the primary spiral lamina (PSL) until contact was obtained as detected by an in-crease in the baseline voltage output of the force transducer. It was then lowered in one-half micron steps at 10 second intervals to a maximum dis-placement of 6-8 microns beyond the pre-loaded state. The voltage output of the transducer was continually recorded. When maximum displacement was reached, the needle was raised, moved 5 microns radially outward, and the process re-peated. Measurement was done until the spiral ligament (SL) was reached, in-volving 26-30 points as the width of the BM in the specimens studied was 135-155 microns. Testing was completed within three hours *post mortem*. Distance from the basal end was determined with a graduated eyepiece.

To eliminate noise, the entire apparatus was situated in an air-tight chamber mounted on a vibration-damping table, and output from the transducer was low-

pass filtered. Since the impedance of the transducer system is infinite, the
distance moved by the transducer needle is exactly the deflection of the BM and
the voltage recorded at each step represents the force necessary to maintain
that deflection.

c. *Calibration*

The sensitivity of the force transducer is 67 mV/dyne as measured with dead-
weight loading. The servo system has a natural frequency of 160 Hz and a
damping ratio of 0.7. Its resolution is .0001%, linearity .05% and non-repeat-
ability .01%, all with regard to full scale. The precision of the entire system was determined to be 3% by testing on a uniform gelatinous block and a
4 micron synthetic plate.

d. *Microstructure*

As an attempt to determine the actual state of the cochlea at the time of force
recording, one cochlea, treated in the same manner as those tested, was fixed
at 4 hours *post mortem* with phosphate-buffered formaldehyde(1%) and glutaral-
dehyde(1%). Post-fixation followed with osmium(2%) in a veronal acetate buffer.
Light microscopic examination of the embedded specimen revealed that the appearance of the BM was identical to its appearance in micrographs of fresh
cochleae. The mesothelial cells lining the BM retained structural integrity
and the inner and outer pillars were intact. Swelling, vacuolization and lysis
were evident in the inner and outer hair cells, Deiter's cells, Claudius's
cells and the cochlear nerve.

3. RESULTS

Graphs displaying force versus BM displacement were obtained for each location

tested, yielding 112 curves for the four animals used in this report. The
curves were trilinear, with the limits of the linear regions varying slightly
from point to point. The first range generally corresponded to displacements

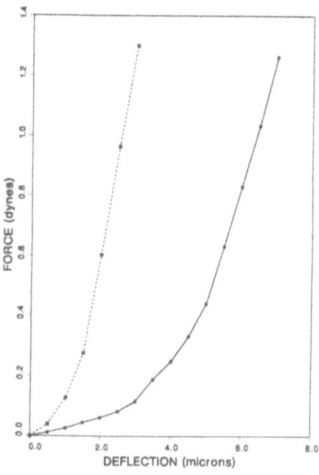

*Fig. 1. Force-deflection curves
for two points on the basilar
membrane.*

of 0-3 microns, the second to 3-5 microns and the third to 5-7 or 8 microns.
For locations near the PSL, however, the extent of the first and second
ranges were very much shorter. Two characteristic curves are shown in Fig. 1.

Figure 2 shows the inverse of the slope of the steepest (third) range of the
force-deflection curves plotted versus radial distance. Each set of points
represents data from one animal. The radial coordinate was normalized with 0.0
at the edge of the PSL and 1.0 at the edge of the SL. Dominant in all curves
is a stiff region extending from the PSL to approximately 0.33. The specimens
in Fig. 2b show a region of intermediate compliance from .33-.56. All
results show localized stiffening at a spot varying from .56-.64.

4. DISCUSSION

From anatomy, it sould appear that the first linear range corresponds to the
loading of the mesothelial cells lining the BM, the second to the amorphous

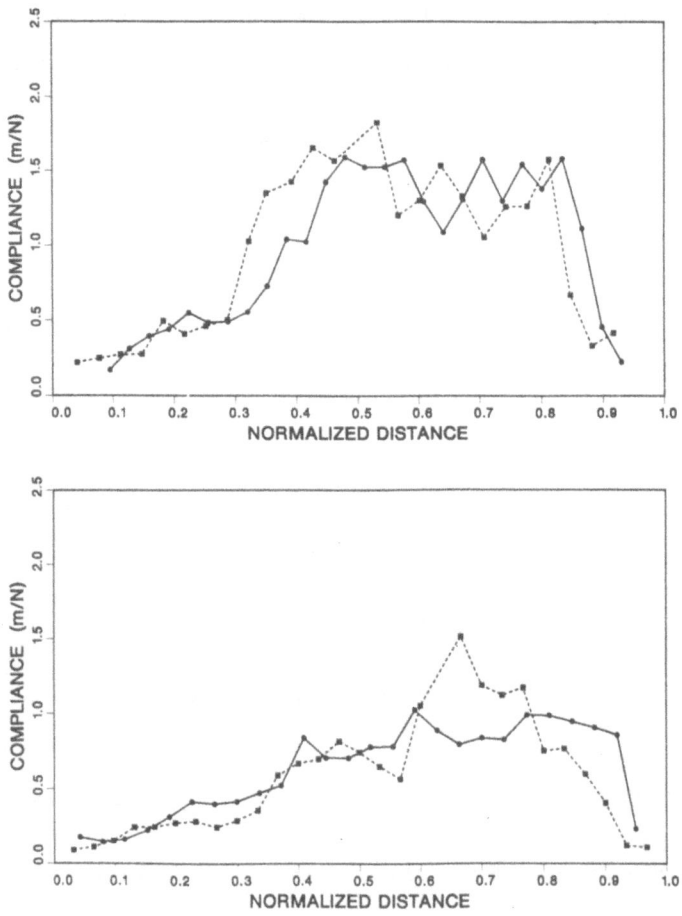

Fig. 2a (top) and 2b (bottom)
Compliance versus radial distance for
four cochleae.

ground substance and the third to the fibers of the BM.   These are the zones

contacted by the needle as it pushes downward from scala tympani.   The arches

of Corti are obviously contributing a stiffening to the structure.   The reasons

for the notch at .56-.64 are as yet unknown; it may be due to support from the

tonofibrils and phalangeal processes of the outer phalangeal (Deiter's) cells.

Although the distance from the stapes was known for each of the radial posi-

tions, no attempt was made to correlate this to compliance since individual variation makes it impossible to compare results from one animal to another. Only one set of measurements was possible per cochlea because of the time required for the procedure.

It is difficult to compute volume compliances from these results, as the deformed shape of the membrane is not known. With the arcuate zone assumed rigid, the present results correspond to equivalent hydrostatic loading volume compliances of .04-.09 mm^4/N. The results of Dancer and Franke (1980) at the 2 mm point are .04 while those of Békésy (1960) for the same point *post mortem* are .18 mm^4/N.

To compare the results obtained here to other measurements of point compliance, we must take into consideration the diameter of the instrument making contact with the fibers. If we assume that there is negligible longitudinal coupling between the fibers of the BM, as demonstrated by Voldřich (1978) for the 15-min. *post mortem* cochlea, then we can model the BM as a set of transverse beams. The needle, of diameter 12.5 microns, should always fully contact at least 5 of the large fibers in the bottom layer of the BM pectinate zone, based on dimensions reported by Iurato (1962). If the diameter of the needle is doubled, the effective moment of inertia of the structures contacted is doubled. If we normalize the present results to an area of fibers one micron in width, the point-load compliances vary from 4 m/N in the arcuate zone to maxima of 25 m/N in the pectinate zone, all steepest range values. The second range gives maxima of 38 and the first range of 63 m/N. The normalized results of Gummer et.al. (1981), who used a 25-micron diameter probe, are then 31-89 m/N for their linear or "plateau" region. The discrepancy may be due to the dynamic nature of the latter's loadings or a difference in the linear zone of partition contacted.

Acknowledgement

This work was partially supported by a grant from the National Science Foundation.

REFERENCES

Békésy, G. (1960). *Experiments in Hearing.* (McGraw-Hill, New York).

Dancer, A. and Franke, R. (1980). "Intracochlear sound pressure measurements in guinea pigs", *Hearing Research 2,* 191-205.

Gummer, A.W., Johnstone, B.M. and Armstrong, N.J. (1981). "Direct measurement of basilar membrane stiffness in the guinea pig", *J. Acoust. Soc. Am. 70,* 1298-1309.

Iurato, S. (1962). "Functional implications of the nature and submicroscopic structure of the tectorial and basilar membranes", *J. Acoust. Soc. Am. 34,* 1386-1395.

Taber, L.A. and Steele, C.R. (1981). "Cochlear model including three-dimensional fluid and four modes of partition flexibility", *J. Acoust. Soc. Am. 70,* 426-436.

Voldřich, L. (1978). "Mechanical properties of the basilar membrane", *Acta Oto-Laryngol. 86,* 331-335.

PROBLEMS IN AURAL SOUND CONDUCTION

L.U.E. Kohllöffel

Institut für Physiologie und Biokybernetik
Erlangen, Germany

ABSTRACT

The ears of many birds and mammals were dissected and examined in search for functional specializations in aural sound conduction. To some of the findings the functional significance is postulated and they are grouped under the headings "The third window in waterbirds" and "Launching the Békésy wave".

1. THE THIRD WINDOW IN WATERBIRDS

In some birds the cochlear aquaeduct is very wide, about as wide as the round window (Werner 1958, 1960). Scala tympani directly faces the brain for about one third to one half of the cochlear length. This "third window" allows rapid pressure exchange between brain and cochlea. After noticing the wide aquaeduct especially in waterbirds - such as penguin, great crested grebe, pelican, flamingo, swan and goose - it occurred to me that the third window may have something to do with underwater acoustic orientation. Underwater directional hearing may be helped by allowing intercochlear fluid flow across the brain.

For the plane underwater sound wave with sound pressure P_F and particle velocity V_F: $P_F = \rho_W c V_F$ (ρ_W mass density of water, c speed of sound in water). The wave affects the ears of submerged birds in two ways. 1. P_F sets the skull into vibration, probably mainly through action on the beak since the head is partly acoustically insulated through the air cushion trapped between feathers. Here we assume that P_F causes the sound pressure P_V in scala vestibuli and P_T in scala tympani. P_T and P_V are assumed equal in both ears and $\Delta P = P_V - P_T$. 2. The bird is carried along by the sound wave with a velocity V_B close to the particle velocity V_F (assume $V_B = V_F$). This is because the mass density of bird ρ_B and water ρ_W is similar (assume $\rho_W = \rho_B = \rho_{BR} = \rho$; ρ_{BR} mass density of brain). Through the acceleration inertial forces are set up in the bird's sound conducting system. Here we ignore all such forces as might arise in the middle and inner ear. We just consider transcerebral flow (Fig. 1a).

The intercochlear pressure P_i arises due to the sound induced velocity V_F of the skull:

$$P_i = j\omega\rho h V_{FX} = j\omega\rho h V_F \cos\alpha = j\omega\frac{h}{c}\cos\alpha P_F \qquad (1.1)$$

($\omega = 2\pi f$; f frequency; V_{FX} component of skull velocity in parallel with inter-cochlear axis X; $\cos\alpha$ direction cosine, h intercochlear distance).

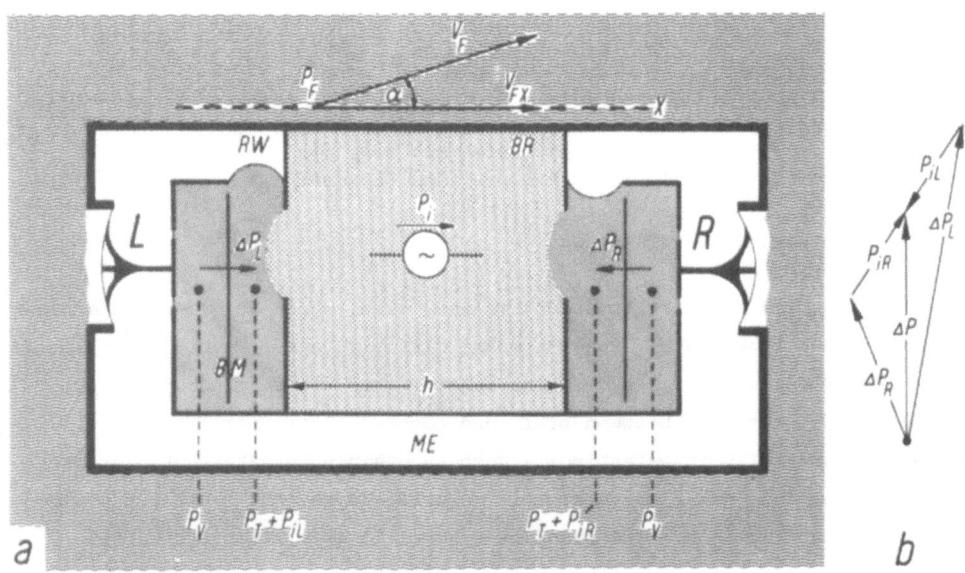

Fig. 1. Scheme for underwater directional hearing in waterbirds. (a): the acoustic field pressure P_F causes sound pressures P_V and P_T ($\Delta P = P_V - P_T$) in scala vestibuli and tympani. Head acceleration causes direction dependent intercochlear pressure P_i and scala tympani pressures P_{iL}, P_{iR}. (b): the left and right (ΔP_L, ΔP_R) transpartitional pressure amplitudes differ and the difference depends on direction of sound wave. ME middle ear space, BR brain, BM basilar membrane, RW round window, L and R left and right cochlea, h intercochlear distance, $V_{FX} = V_F \cos\alpha$, V_F particle velocity, α angle between direction of sound wave and intercochlear axis X.

P_i is an open circuit pressure; it could be measured across the brain if either one or both aquaeducts were rigidly closed so blocking intercochlear flow. P_i drives intercochlear volume flow against the internal (intracerebral) acoustic impedance Z_i and the acoustic impedances at the two third windows ($2Z_{3.W}$). If most of the flow is shunted through the round window impedances (Z_{RW}) it follows that $2Z_{3.W} = 2Z_{RW}$. P_i causes the pressure P_{iL} in the left and P_{iR} in the right scala tympani:

$$P_{iL} = -P_{iR} = \frac{Z_{RW}}{2Z_{RW}+Z_i} P_i \qquad (1.2)$$

For the pressure across the cochlear partition in the left (ΔP_L) and the right (ΔP_R) ear:

$$\Delta P_L = \Delta P - P_{iL}, \quad \Delta P_R = \Delta P - P_{iR}, \quad \Delta P = P_V - P_T$$

$$\Delta P_L - \Delta P_R = -\frac{2Z_{RW}}{2Z_{RW} + Z_i} \, j\omega\frac{h}{c}\cos\alpha P_F$$

$$(1.3)$$

Thus the intercochlear flow causes a difference in amplitude between the left and right transpartitional pressure (Fig. 1b). The difference depends on the direction of the sound wave. Underwater directional hearing of waterbirds may operate on this principle.

2. LAUNCHING THE BEKESY WAVE

The delivery of power to the acoustic load Z_B by an acoustic system can be described as shown in Fig. 2.a. P_0 is the sound pressure between d and e for $Z_B = \infty$, Z_{in} is the internal impedance. At the circular frequency ω_y maximum real power is absorbed by $Z_B(\omega_y)$ if for the real (Re) and imaginary (Im) parts:

$$\text{Re}Z_B(\omega_y) = \text{Re}Z_{in}(\omega_y), \quad \text{Im}Z_B(\omega_y) = -\text{Im}Z_{in}(\omega_y) \qquad (2.1)$$

Figure 2b: Let $Z_B = R_B$, $Z_{in} = R_{in} + (j\omega C_{in})^{-1}$. The power $|P_B|^2 R_B^{-1}$ absorbed by R_B can be changed by shunting R_B with an acoustic mass L_S and so forming a high pass filter. At low frequencies inserting L_S causes an insertion loss. However, it follows from Eqs. (2.1) that there is an insertion gain IG(ω_B) at ω_B if:

$$L_S = C_{in}R_{in}R_B, \quad \omega_B = C^{-1}\{R_{in}(R_B - R_{in})\}^{-\frac{1}{2}}, \quad R_B > R_{in}$$

$$IG(\omega_B) = 10 \log \frac{\text{Power absorbed with shunt}}{\text{Power absorbed without shunt}} = 10 \log \frac{R_B + 3R_{in}}{4R_{in}}$$

$$(2.2)$$

If $R_B = 9R_{in}$ the shunted R_B absorbs three times as much power as the non-shunted R_B at ω_B.

Figure 2c: Let $Z_B = R_B$, $Z_{in} = R_{in} + j\omega L_{in}$. Power absorption can be changed by inserting the acoustic shunt compliance C_S so forming a low pass filter. At high frequencies one obtains an insertion loss. Using Eqs. (2.1) one obtains an insertion gain at ω_c if:

$$C_S = L_{in}(R_{in}R_B)^{-1}, \quad \omega_c = L_{in}^{-1}\{R_{in}(R_B-R_{in})\}^{\frac{1}{2}}, \quad R_B > R_{in}$$

$$IG(\omega_c) = 10 \log \frac{R_B + 3R_{in}}{4R_{in}}$$

(2.3)

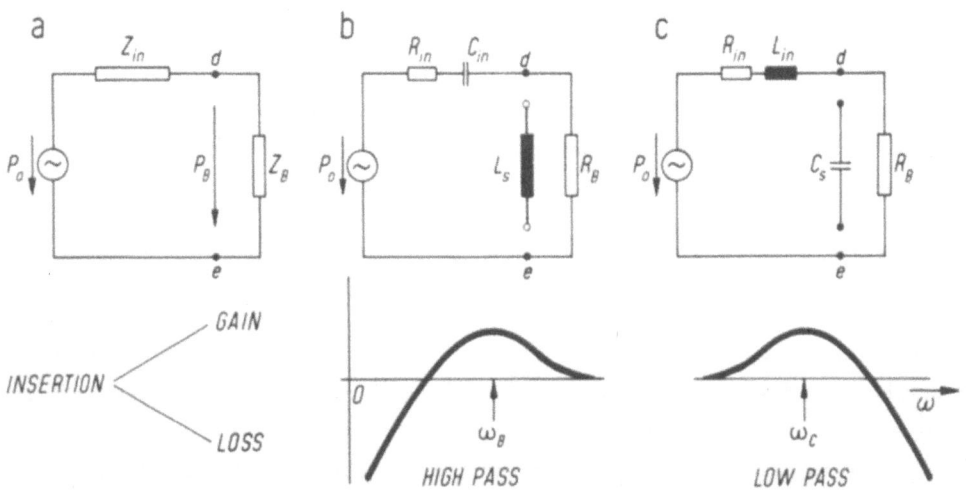

Fig. 2. Power matched sound transmission to the Békésy wave.
(b,c): depending on internal impedance Z_{in} power absorption in a
certain frequency band can be improved by shunting wave resistance
R_B with acoustic reactance.

The scheme may be applied to the cochlea. Z_B is the input impedance of the
Békésy wave that transports power into the cochlea; $Z_B = R_B$ in the frequency
range without reflections from the helicotrema. Z_{in} is the source impedance
at the fenestral boundaries. It is composed of the acoustic mass due to the
flow field that couples wave- and boundary motion, plus the acoustic impedance at round and oval window. The latter contains the middle ear impedance
and the radiation impedance of the eardrum as seen at the stapes footplate.
P_o is the transpartitional pressure due to the external sound field if
$Z_B = \infty$.

At frequencies below middle ear resonance $Z_{in} = R_{in} + (j\omega C_{in})^{-1}$. Power absorption by the Békésy wave can be improved in a certain frequency range by
shunting R_B with an acoustic mass.

a. The Ductus Brevis

The ductus brevis in birds was discovered by de Burlet (1929, 1934). It can
be seen in drawings of the parrot cochlea by Denker (1907), but Denker did

not *recognize* the duct. The duct runs between the neural limbus and the bony wall. In contrast to the helicotrema it interconnects the cochlear perilymphatic canals at the *basal* end. Beyond a certain cut off frequency the duct presents a pure acoustic mass that shunts the cochlear partition at the basal end.

The basal situation differs between birds. The ductus brevis is absent in owls (Schwartzkopff and Winter 1960), it is extremely narrow – nonfunctional – in chicken birds such as turkey, pheasant, quail, it is present in pigeon, songbirds, woodpecker, duck, and it is particularly wide in goose (duct in goose: diameter 0.6mm, length 2mm).

At low frequencies the ductus brevis controls transpartitional pressure and it causes an insertion loss. However, in a certain frequency band below middle ear resonance it can cause an insertion gain (Eqs. (2.2), Fig. 2b).

b. The Lamina Problem

The problem of the flexible lamina spiralis ossea primaria (spl) is old (Perrault 1680, DuVerney 1684) yet few workers in hearing are aware that it exists (Steele 1976, Taber and Steele 1981). The spl is part of the cochlear partition. It is wide in the basal turn where the basilar membrane (bm) is narrow. It narrows towards the apex as the bm widens. In many mammals spl is thick and rigid and it moves much less than the bm. These include the laboratory mammals used in auditory physiology: cat, rat, chinchilla, guinea pig, mongolian gerbil, squirrel monkey, rabbit. In some mammals the basal spl (bspl) is very thin, fragile and flexible: for instance in pig, cow, man and mole (in mole just for the most basal 2mm). In pig and cow bspl deflects almost as much as the basal bm (bbm) at low frequencies, and in case the bm is stiffened by formalin fixation bspl deflects more than bbm. In unfixed human preparations (2 days postmortem) bspl and bbm deflect with equal amplitude over the distance from 3 to 14 mm at sound frequencies up to 1 kHz. Under a static point load (hair probe) the human bspl deflects nearly as much as the round window membrane. Polvogt and Crowe (1937) reported that a partly deossified bspl is compatible with normal hearing in man.

What is the functional significance of the flexible bspl? (It is probably not just the lack of sensitivity to ultrasound that allows an ear with flexible lamina.) The analogy to the acoustic plate absorber might be useful. At high frequencies the wavelength is short and the transpartitional pressure wave has a steep longitudinal gradient. So bspl is stiff owing to its longitudinal bending stiffness. At low frequencies the wavelength is long and the

transpartitional pressure distribution approximates to a spatially uniform
pressure load. To this uniform load bspl yields like a soft plate since only
the small radial bending stiffness provides restoring forces. Here bspl may
have a **resonance** and beyond the resonance frequency its motion may be mass
controlled provided the pressure load remains sufficiently uniform. Thus in a
certain frequency band the flexible lamina may have an effect similar to
shunting the Békésy wave with a basal mass reactance. If this happened below
middle ear resonance power absorption by R_B could be improved (Eqs. (2.2),
Fig. 2b).

At frequencies above middle ear resonance $Z_{in} = R_{in} + j\omega L_{in}$. Part of the internal acoustic mass L_{in} is due to the inertial flow field that couples footplate
motion to wave motion. In noctural owls (Strix aluco, Strix flammea) the bony
frame of the oval window projects deeply in between the cochlear and vestibular fluid compartments and the columella footplate acquires as a result the
striking drop-like profile (Krause 1901, de Burlet 1934 Fig. 1238). This separation between cochlea and vestibule changes the flow field. It may reduce
the mass load imposed by the flow field and may thus improve high frequency
sensitivity. Generally, power absorption can be improved by shunting R_B with
an acoustic compliance (Eqs. (2.3), Fig. 2c).

c. The Limbus Problem

In birds the bm is carried between the neural and abneural branch of the
limbus. Over the basal part of the cochlea (along the cochlear windows) the
limbus branches are only loosely anchored at the cochlear wall (independent
of the presence or absence of the ductus brevis). Under a static point load
they yield like elastic cantilevers supported in the upper cochlea. They do
so too under an alternating pressure load. This seems to be the case also
in caiman (Smolders, Wilson and Klinke 1982). The vibrating basal limbus
could act similar to a compliant shunt of the basilar membrane at the basal
end, thereby improving power absorption by R_B at frequencies above middle ear
resonance (Eqs. (2.3), Fig. 2c).

REFERENCES

De Burlet, H.M. (1929). Zur vergleichenden Anatomie und Physiologie des peri-
    lymphatischen Raumes, *Acta Oto-Laryngol.* 13, 153-187.
De Burlet, H.M. (1934). Vergleichende Anatomie des stato-akustischen Organs.
    In: *Handbuch der vergleichenden Anatomie der Wirbeltiere* (Urban und
    Schwarzenberg, Berlin), Vol. 2, pp. 1293-1432.

Denker, A. (1907). *Das Gehörorgan und die Sprechwerkzeuge der Papageien* (Bergmann, Wiesbaden).

DuVerney, G.J. (1684). *Tractatus de organo auditus* (Norimbergae).

Krause, G. (1901). *Die Columella der Vögel (Columella Auris Avium)* (R. Friedländer und Sohn, Berlin).

Perrault, Cl. (1680). *Essais de Physique* (Coignard, Paris).

Polvogt, L.M. and Crowe, S.J. (1937). Anomalies of the cochlea in patients with normal hearing, *Ann. Otol. Rhinol. Laryngol.* 46, 579-591.

Schwartzkopff, J. and Winter, P. (1960). Zur Anatomie der Vogel-Cochlea unter natürlichen Bedingungen, *Biol. Zbl.* 79, 607-625.

Smolders, J., Wilson, J.R., and Klinke, R. (1982). Mechanical frequency analysis in the cochlear duct of Caiman Crocodilus, *Pflüg. Arch.* Suppl. 392, R51.

Steele, C.R. (1976). Cochlear mechanics. In: *Handbook of Sensory Physiology*, edited by W.D. Keidel and W.D. Neff (Springer, Berlin-Heidelberg-New York), Vol. 5, Part 3, pp. 443-478.

Taber, L.A. and Steele, C.R. (1981). Cochlear model including three-dimensional fluid and four modes of partition flexibility, *J. Acoust. Soc. Amer.* 70, (2), 426-436.

Werner, Cl. F. (1958). Der Canaliculus (Aquaeductus) cochleae und seine Beziehungen zu den Kanälen des 9. und 10. Hirnnerven bei den Vögeln, *Zool. Jahrb.* 77, 1-8.

Werner, Cl. F. (1960). *Das Gehörorgan der Wirbeltiere und des Menschen* (Thieme Leipzig)